City Leaders for Sustainable Futures

Cathy Oke • Jorn Verbeeck

City Leaders for Sustainable Futures

Portraits of Visionary Urban Changemakers

palgrave macmillan

Cathy Oke
Melbourne Centre for Cities
University of Melbourne
Parkville, VIC, Australia

Jorn Verbeeck
Dark Matter Labs
Kontich, Belgium

ISBN 978-981-95-9543-3 ISBN 978-981-95-9544-0 (eBook)
https://doi.org/10.1007/978-981-95-9544-0

This Palgrave Macmillan imprint is published by the registered company Springer Nature Singapore Pte Ltd.
The registered company address is: 152 Beach Road, #21-01/04 Gateway East, Singapore 189721, Singapore

Foreword

After collective decades working with urban leaders across continents—from community organisers in informal settlements to researchers and policymakers bridging science and practice—I have learned that alongside figuring out *what* cities should do about climate change, we need to be thinking about *who* drives transformation and what sustains them through years of intense and difficult work.

This book answers that question with rare honesty. Twenty portraits reveal the real lives of people helping shape, guide, and cultivate cities from multiple vantage points—Indigenous activists, young climate advocates, network builders, and a few elected officials. It also features two portraits of authors Cathy and Jorn—friends first but city-shaping boundary spanners in their own right who carry a wealth of experience at the crossroads of science, policy, and practice. Far from a collective vanity piece, these portraits tell a story of perseverance, vision, and relentlessness in the face of rapidly shifting priorities and demands.

Working globally with the Global Covenant of Mayors for Climate and Energy alliance, I have witnessed how urban transformation can emerge from diverse actors operating across scope and scale—not just from city halls but from neighbourhoods, networks, and movements. The leaders profiled here show what sustained commitment looks like, rooted in both the personal and the institutional.

For anyone working in urban development, climate action, or governance, this collection offers something essential: evidence that transformation requires both technical competence and emotional intelligence and strategic thinking and cultural wisdom. These stories provide compass points for navigating uncertainty. In revealing how these 20 individuals found their paths, Jorn and Cathy help readers discover their own.

Benjamin Jance IV
Director, Climate Action
Innovation, Research, and Impact,
Global Covenant of Mayors for Climate
and Energy (GCoM) Secretariat
Brussels, Belgium

Preface

Behind any book lies a community of believers. Our community includes 20 remarkable individuals who trusted us with their personal odysseys. These urban leaders opened their lives to us, revealing not so much their professional achievements as the emotional and spiritual dimensions of their work. Their willingness to be vulnerable, to share failures alongside successes, transformed what might have been a conventional collection of profiles on leadership into something richer: a kaleidoscope of possibility for how we might reimagine our urban futures. We thank them all profoundly.

This book deliberately departs from conventional urban studies literature. We have chosen to explore the lived experiences of those committed to collective well-being. The portraits treat cities not as mechanical systems requiring optimisation but as living contexts where personal encounters with displacement, belonging, and discovery translate into collective action. These are stories of long-term weavers—people who understand that creating sustainable futures requires patience, cultural intelligence, and the courage to privilege care over capital. We offer these narratives as inspiration and invitation to urban practitioners seeking to understand how change unfolds beyond spreadsheets and strategic plans; to students discovering that technical expertise alone cannot address our greatest challenges; and to engaged citizens recognising that the future of cities depends on understanding both sophisticated solutions and the deeply human commitments that sustain them over decades. This book is also for national leaders exploring impact looking holistically at community beyond ideological identity, and it is for entrepreneurs anchoring creativity and innovation into the urban fabric that often offers the key conditions for value to be created.

We are grateful to the Faber Academy Narrative Non-Fiction class of 2023/2024, who helped us discover and clarify our writing voice during the crucial early stages. Ellen O'Brien's edit of an early story extract showed us how these narratives could breathe on the page. Rob Oke, Kay Oke, Chris Cane and Lara Piret provided encouragement and technical review at moments when we questioned whether personal stories could carry the weight of urgent urban challenges. Thanks to the anonymous reviewers of this manuscript whose feedback completed the book's final form and to Laura Goodin for copy-editing.

Throughout the writing process, Charley, Yann, Viggo, Astor, and Alix provided the familial support that all creative work requires: patience when deadlines loomed, curiosity about the emerging narratives, and the grounding presence that reminds us why sustainable cities matter in the first place.

May these stories help you find your own pathway towards the urban futures we desperately need—and deserve—to create together.

Parkville, VIC, Australia Cathy Oke
Kontich, Belgium Jorn Verbeeck

About this Book

This book on cities reveals pockets of optimism among a diverse group of individuals navigating complexity, spanning boundaries, building coalitions, and sustaining change across political cycles. Cathy Oke and Jorn Verbeeck present 20 intimate portraits where personal endeavours and systemic change intertwine. Urban leaders like Mark Watts, Maimunah Mohd Sharif, Carlos Moreno, Jess Miller, Marvin Rees, Sheela Patel, and Debra Roberts share about their journeys in embracing living systems. From Indigenous wisdom keepers to systems innovators, these boundary spanners reveal that sustainable cities and communities emerge not through perfect plans but via patient relationship building and cultural wisdom that cross generations. Rather than celebrating lone heroes, the book reveals cities as 'kaleidoscopes of possibility' that require ensemble casts who translate between disciplines, sectors, and communities. Each story weaves personal odyssey with professional transformation, inviting readers to reflect on their own moments of inspiration and pathways to change and offer a soft framework and gentle invitation to share their inspiration with others.

This is a syllabus for educational and learning approaches to sustainable city leadership that integrate multiple knowledge systems, honour ancient wisdom alongside contemporary innovation, and foster collaborative intelligence and community building by the stories we share. Essential reading for anyone navigating the messy, vital work of making cities thrive.

Contents

About the Author

Cathy Oke (Australia) OAM is Director of the Melbourne Centre for Cities and Associate Professor in Informed Cities at the Faculty of Architecture, Building and Planning at the University of Melbourne. She is a Lead Author for the IPCC Special Report on Climate Change and Cities and Senior Advisor to the Global Covenant of Mayors Innovate4Cities initiative. Cathy was a Councillor at the City of Melbourne from 2008 to 2020. In 2023, she was awarded an Order of Australia Medal (OAM) for service to local government and the environment. She loves sharing her passion for cities, nature, and music with her family, especially daughter Charley and husband Chris.

Jorn Verbeeck (Belgium) is an Urban Strategist with Dark Matter Labs, based in Belgium. A journalist by training with over 20 years of sustainability expertise across public and private sectors at local and international levels, he specialises in urban challenges, viewing cities as the daily negotiation between societal actors and at the crossroads of societal transitions. His expertise spans city diplomacy, sustainability strategies, multi-level governance, climate action and investment plans, circular economies, environmental, social, and governance assessments, and smart city approaches. He serves as a Board Member for the EU Climate Neutral and Smart Cities Mission and Leading Cities, and has previously worked with the City of Antwerp and Ghent, and the Global Covenant of Mayors for Climate and Energy.

Chris Cane (Australia) is a high school art and music teacher with degrees in industrial design, education, and music performance. An accomplished and diverse educator, his passions are live music, playing guitar, and gardening. His favourite city is Stockholm but loves his hometown, Melbourne.

List of Figures

1

A Syllabus for Sustainable City Leaders

Abstract This book presents 20 portraits of urban leaders whose personal journeys, from our perspective, illuminate a possible syllabus for sustainable city futures. These 20 personal stories are but a representation of conversations we have had with many inspirational city leaders over the years, that have stuck with us long after they occurred. We share these loosely structured pieces, gathered from 60- minute interviews, letting our interviewees responses to provide direction, leaving thoughts unfinished and some topics and questions unanswered.

Keywords Boundary spanning • Transformative leadership • Urban governance • Capacity building • Systems thinking

Embracing the Urban Kaleidoscope

There is something about cities, big and small. Places most of humankind calls home, whether formal or informal. Characterised by something profoundly human. Spaces that blend the individual, collective, and relational; the personal and professional; the planned and nascent; the factual

C. Oke, J. Verbeeck, *City Leaders for Sustainable Futures*,
https://doi.org/10.1007/978-981-95-9544-0_1

and imaginary; the bright lights and twilight; the pockets of the future and testimonials of the past. Systems that over time search, struggle, thrive, and decay. Urban metabolisms that invisibly curate networks of interdependencies.

The idea for this book probably started during one of our very first conversations in the late spring of 2020, having just become colleagues in a global city network. From the opposite sides of the globe, having both worked in many different urban positions, we reflected on what makes up urban life and to what extent urban challenges and transitions can be managed. From Antwerp to Melbourne, some characteristics were very much contextual, but others appeared deeply universal. Our conversations, it seemed, often centred not about the bricks and mortar but rather about how cities are being governed, how they build capacity to support change and resilience, and about the intersections of urban challenges and opportunities and the people working at them: the urban fabric between institutions and lived reality; between the 'makeable' city and its policies, but at the same time also the urban culture, spontaneity and informality.

With the urban kaleidoscope comes an almost by-design paradox. The city as a place of both (be)longing and unbelonging. Human settlements are often characterised by a healthy friction of relationships, revolution, and resilience, on the one hand, and anonymity, alienation, exclusion, and inequality, on the other. Whether or not it is this contradiction that helps drive transformation, urban environments are change agents in themselves. Despite or because of being in the front line against many grand societal challenges, cities somehow seem to safeguard our hope for the future or provide at least some form of stability in a fast-changing world. Because of a familiarity, legibility, and spirit of a place? Or by a form of fairness, a sense of what the human collective can achieve? Though having always been in constant flux, over the past centuries many cities have witnessed and have seen urban growth accelerate and with it urban complexity. Working in urban development has therefore become less a matter of solutions and more a matter of decision-making systems that translate complexity into a portfolio of choices. Attempts to replicate best practices are being replaced by initiatives that help build trust and relationality between people who understand how cities actually function, rather than how they should function. Traditional urban planning geared towards efficiency is evolving towards systems design that recognises how cities relate the individual to the collective by providing a physical and social environment, and towards shaping enablers for communities where personal lives and experiences merge into a shared urban dynamic.

Cities Are Shaped by Their Stories

Cities have always been somehow imperfect, but what if change is happening faster than humans can adapt? One might say that climate change currently amplifies inequality; that housing insecurity amplifies political anger; that flooding amplifies mistrust. True or not, correlations prove to be tricky, and the current global order and nation-states struggle to come up with a convincing narrative. As it is often in human settlements that crises collapse into each other, can urban environments find ways to address these many wicked problems and offer tangible pathways to resilience? The answer is brutally conditional. Cities can concentrate creativity, innovation, capacity, and cultural change, but poorly managed urban development locks in fragmentation, sprawl, pollution, emissions, and inequality. Despite an overwhelming number of good practices and initiatives, for various reasons, many cities seem stuck in a loop of pilots and plans, fragmented agendas, and frictional politics. This is not a judgement but a continuous conversation between many of the people we work with, who share a quiet but firm conviction: that sustainability and resilience will stand or fall in cities. How can we therefore create the conditions for the speed and scale needed to establish healthy, clean, and resilient living environments? As urban growth trajectories will shape the demand for energy and resources and the well-being of billions, the key question is how to orchestrate governance, institutions, and capacity to plan and implement.

Cities are often described as engines—of growth, innovation, and culture. But an engine implies design, control, and ownership. A city is not an engine. It is an everyday negotiation between people and land; past and future; money, materiality, and immateriality; security and precarity.

It is a negotiation between scales—personal and professional, household and neighbourhood, municipal and national, local and global—and between the city, its bioregion and its more-than-human elements: rivers, soils, trees, coastlines, pollinators, and tides. Cities tackle crises not by being cities but by functioning as catalysts, becoming governable in new ways through deepened relationality and healthy entanglement.

What does it means in practice? Here is the provocation: the limiting factor is often not imagination nor technology. It is mostly about leadership capacity, not in the sense of charismatic individuals but in the sense of living ecosystems that are able to hold complexity and choices, translate them into engaging stories, and manage friction. And it is not about mayors alone nor solely about planners, activists, social workers, community leaders, or entrepreneurs. It is about all of them and the commitment and respect they hold between them. Much of what makes a city are its change agents who can span boundaries, build coalitions, and make change durable across time by the stories they tell. When they amplify the voices and initiatives of both people and place, their stories are a crucial, invisible infrastructure in urban development. They provide a human-centric way to connect the tangible with the intangible, test new paradigms, enable collective movement, and shape a shared vision and identity for the urban and societal future.

When we talk about 'urban leadership,' we mean participating in this negotiation, trying to shift it towards stewardship. That shift is never neutral. It is shaped by power. Territory matters as the terrain through which global relationships become real. So who owns the city? Jane Jacobs insisted that cities are made through everyday life, through the 'sidewalk ballet' of eyes and encounters.[1] 'Ownership' often is about legitimacy: who decides what the city is for, whose needs are strategic, and whose are noise. There Indigenous perspectives offer powerful correctives, among which is the notion of 'country.' Country is not a backdrop but a relation.[2] Listening is not consultation but ethical posture. This shifts the question from 'How do we get buy-in?' to 'How do we become

[1] Jacobs, J. (1961). *The death and life of great American cities* (pp. 50–54). Random House.

[2] Nicholson, M. S. (2023). On Country off Country. Monash University. https://doi.org/10.26180/24707391.v1

accountable?'. It also shifts time horizons: decisions are contributions to long stories of stewardship. It furthermore changes what counts as competence: not only accelerating delivery but every now and then also slowing down to listen, connect, reorient, and repair.

Leadership and Boundary Spanning Are Crafts, Not Traits

Let us be blunt: sustainable future enthusiasts have perhaps overinvested in two myths. The first myth is the technical myth: if we get the data, the tools, the policy design, and the finance mechanism, then the city will transition towards urban resilience. The second myth is the hero myth: if we get the right leader, the visionary mayor, the genius entrepreneur, and the charismatic activist, then the city and maybe even the wider society will transition too. Both myths ignore that aspect of relationality. Relationships are messy. They are slow. They are full of misrecognition. They are shaped by history. They can be tender and violent at the same time. And yet, they are the only thing that can hold the city together long enough for change to become durable.

This is why boundary spanning matters. It is the lived work of moving between worlds: between disciplines, between sectors, between communities, between political realities, and between evidence and emotion. It is also often invisible work, hard to measure, hard to reward, and easy to burn out on. Boundary spanners act as catalysts, intermediaries, and brokers and help create enabling conditions, build trust, translate concepts, and make climate considerations feel like a natural part of governance rather than an external imposition.[3] The work resists straightforward quantification because impacts unfold through long, iterative interactions. Climate change, migration, ageing, and inequality do not care about election cycles. Children do not grow on budget cycles. Rivers do not heal on quarterly reporting. Yet institutions behave as if the future

[3] Moosavi, S., Keane, B., Roberts, D. et al. Tales of boundary spanning for climate actions in cities: from theory to practice, and back. *npj Urban Sustain* 5, 59 (2025). https://doi.org/10.1038/s42949-025-00246-4

can be deferred. This asks a different question: what future are we training ourselves to make? It expands the leadership's audience. The leader is accountable not only to today's voters but to future residents who cannot yet speak, to more-than-human worlds that do not vote but do sustain life.

Maybe we have created a global society too complex to thrive. We have more information than ever, yet moral compasses lag. Grand narratives and modern mythologies that cut through noise and offer belonging are lacking. If investment optimises short-term returns, cities produce fragility despite claiming resilience. So perhaps urban leadership is the capacity to do boundary work without losing the soul. It is the ability to stand with one foot in an institution and one foot in the street and to remain credible to both. It is the ability to negotiate between urgency and patience. It is the ability to carry conflict and multiple truths without turning them into polarisation and simplification.

To come back to the provocation: cities should not plan better but orchestrate differently. Skills and capabilities are uneven. Leadership must join up partnership building, regulation, implementation, and finance, not separate them into disconnected streams. How can leaders move from acting as planning authorities to functioning as learning ecosystems across sectors, cultures, and time? How can leaders move beyond narratives that talk about 'scaling solutions' without asking who bears the cost, who generates the multiple forms of value, and who defines success? A small thought bulb that started this book was asking each other—can we, by giving a voice to the people that inspire us, shape a language and framework to help grow the community of boundary spanners?

About This Book

Years of working in cities have given us something more than expertise: a recurring experience of conversations that started somewhere specific and ended somewhere unexpected. Topical discussions that drifted into meetings of minds. Fields overlapping. Personal stories intersecting with professional visions. Futures becoming imaginable in the middle of a sentence.

Much has been written about cities already, but we wanted to document a small sample of these agents of change we have been lucky to witness as friends or colleagues. If we are right that boundary spanning is underexplored or underinvested in as a pathway for sustainability, we felt that we could contribute a book on cities in which we could share a glimpse of dialogues we have had over the past years.

This book presents 20 intimate portraits of urban leaders whose personal journeys, from our perspective, illuminate possible pathways towards sustainable city futures. These 20 are but a representation of conversations that have stuck with us long after they occurred. We share stories that are loosely structured, letting interviewees provide direction, leaving thoughts unfinished and some topics and questions unanswered. Each chapter presents a short essay that preserves each participant's authentic voice, crafted from interviews over the past 2–3 years. The participants' work and stories are not just a collection of good practices; they also offer a form of long-term perspective and quiet mentorship that can perhaps help provide inspiration on how to navigate change and uncertainty. Their boundary spanning work is opening a conversation space. The book is not designed to be read sequentially. It is a multi-layered invitation, where we hope you will reach out to us to help build the field of boundary spanning and the many intersections it holds.

We do not know yet where it will take us. We explore whether we should perhaps shape a proper school of boundary spanning: not a soft add-on or conference circuit but a shared civic responsibility, a shared caring, as a form of governance. New generations step up not by being thrown into the fire but inducted into craft through working with mentors and peers, taking time to reflect, and having permission to be uncertain. The urban transition is not only technical and political but also very much personal, moving through hopes and dreams alongside fear and failure. The strongest leaders do not pretend fearlessness but build structures that let courage and fragility become collective.

Guardians of the Future, a Syllabus for Hope

Imagine the participants gathered, not in conference halls but in spaces that matter to them. Debra Roberts beneath her acacia in Zimbabwe's savannah. Winston Chow along Bangkok's Chao Phraya River. Cynthia Houniuhi under a Vanuatu banyan tree asking if future generations will see the same beauty. Sophie Howe on Oxwich Beach where childhood memories anchor Wales's Future Generations legislation. Maimunah Mohd Sharif by Malaysian rivers bringing village wisdom to global urban policy.

They range from Indigenous knowledge keepers to former mayors and institutional architects, academics and systems scientists to CEOs and creative practitioners, and innovation catalysts to community organisers. They speak to us from Durban to Paris, Melbourne to Bangkok, Bristol to Moca. They are 'long-term weavers' whose intimate stories demonstrate how personal encounters with displacement, discovery, and belonging shape approaches to collective challenges. All, in a way, share about the shift from lone heroes to ensemble casts and the dream team they themselves would perhaps gather.

Their stories reveal a new narrative emerging: one that prioritises care over capital, climate resilience over risk, and collective well-being over individual advancement. These stories help build the invisible architecture and infrastructure of cities. Read this book, therefore, with hope. Read it with optimism born from the diversity of paths taken and approaches to resilience emerging from pockets of the world. These 20 odysseys constitute a syllabus for future cities—not a prescriptive curriculum but an invitation to possibility.

So, here is the invitation: if you work in cities, do not only ask: What is my project? Also ask: What is my boundary? What is the community I am serving? The how that is underdocumented and that I should share? Where do I have the legitimacy, access, and skill to bridge two worlds that currently do not speak? Where can I translate without diluting? Where can I convene without controlling? Where can I protect builders? Where can I make time visible? Where can I carry a future audience into today's decisions? And as the work is far bigger than any one

Fig. 1.1 Jorn Verbeeck and Cathy Oke, drawing of authors' own selfie

organisation, then also ask: Who else needs to be in the room for this to become real? What is the dream team it takes to uphold the urban, climate, and common future quest?

These are not rhetorical questions but operational ones. They drive how every guardian in these pages navigates their particular terrain between what is and what might be. The question is not whether you will join them. You already have, if you are reading this. The question is how you will weave your thread into the larger tapestry, what pattern you will help create, whether you will create a feedback loop to us, and whose story will intersect with yours in the great urban kaleidoscope of our becoming (Fig. 1.1).

Cathy and Jorn
February 2026

2

Meredith Adler (Canada): The Space Between Foundations and Builders

Abstract Meredith is a climate and energy strategist working at the intersection of cleantech, utilities, and communities. She spent a decade building Student Energy, a global charity that enables young people to build energy careers, and advocated for students' crucial role within climate initiatives. Fascinated by challenges in deploying climate solutions, Meredith has moved on to work in novel cleantech, both with energy utilities and in partnership with Indigenous communities.

Keywords Energy transition • Youth leadership • Capacity building

30 January 2024, Online from Squamish, British Columbia, Canada

Against the backdrop of Pacific Northwest winter light, Meredith Adler (Fig. 2.1) calls in from her small town, just outside Vancouver, having recently transitioned from nearly a decade leading Student Energy to exploring the industrial cleantech world.

C. Oke, J. Verbeeck, *City Leaders for Sustainable Futures*,
https://doi.org/10.1007/978-981-95-9544-0_2

Fig. 2.1 Meredith Adler

She does not have to think long when asked where she would take us if not online. She would transport us not to any contemporary city but to San Francisco circa 2004—a place and time that for her capture something essential about urban possibility that she fears we may be losing.

'At the time, you could walk to different art spaces and record stores. This was before children having cell phones or anything like that. But my parents were totally fine just letting me kind of do whatever I wanted all day, and then showing up at a meeting spot at the end of the day. It was a very inspiring time for me, with a lot of independence and a lot of freedom, where the city allowed you to really get to know different people. That was very unique. It's almost a point of sadness for me to have these memories, because now I wonder whether a rational parent would let their 13–14-year-olds wander just by themselves. My parents were a little on the cutting edge of free-range parenting, I suppose.'

It is not mere nostalgia but a deliberate return to a moment when cities felt genuinely accessible to young people seeking to discover themselves and their place in the world. She muses how San Francisco nurtured discovery and identity formation, when she was reading Kerouac and being obsessed with neighbourhoods like Haight-Ashbury. The record stores of that era provided particularly powerful experiences. 'I was very into punk rock at the time and ska music; the likes of Pennywis then often played in the local shops. People still make fun of me, but I think their first album still is one of my favourite albums. It was very influential—you both know that I ended up challenging a lot of authority in my life, so I think it's still a good soundtrack to my journey,' she recalls with a touch of humour.

'Many of my favourite moments finding my way were just feeling like you could kind of get into the ebb and flow of a city and kind of belong, which sometimes happens very quickly within some cities. How can we start to replicate that? A lot of my inspiration has been around the intersection of inequality and climate work. Cities, I believe, are really the hub of that. I think the disconnect starts when people in power have too much distance from the land they're on, the communities they're in, or the communities that they're impacting. That is when a sense of humanity gets lost. Now working in industrial cleantech, I see investors and other actors who are just constantly thinking about the silver bullet of technology. But if you just do technology, you end up with a few people doing well at the top, with rapid inequality, with systems that don't work for everyone. There has to be an eye on the prize of a holistic transition, rather than just thinking about one solution in isolation.'

That teenage experience of urban freedom created a template for understanding what cities can provide: 'At their best, cities really have the opportunity to make life easier,' she reflects. 'Within a city, you can have your amenities, childcare, all those sorts of things within walking distance, or easy public transport.' This is not just about convenience but about the kind of life such accessibility enables, where 'people do have better quality of life, versus cities where you're stuck in traffic for four hours a day—that's four hours that you're not either working, being with your family, or doing something more meaningful for you.'

Fig. 2.2 A report, illustrative of the object mentioned by Meredith Adler

The object Meredith chooses to represent her journey shows how personal and political transformation intertwined during her university years. Rather than a physical artifact, she describes an academic paper on water governance that crystallised her understanding of how social justice and environmental challenges connect. 'I actually came to sustainability fairly late in life compared to a lot of my peers,' she explains, describing how a year living in Argentina after high school taught her about Indigenous rights and exposed gaps in her American education (Fig. 2.2).

Initially she really saw [Indigenous rights] as totally devoid from sustainability, 'because most of what I knew about sustainability was really conservation work.' The apparent contradiction troubled her: 'Many Indigenous people have been disenfranchised by conservation efforts, making sustainability and my human-centred values seem fundamentally misaligned.' University courses on food systems and water governance led to a transformational insight: 'You can donate however much money you want to human rights, but if you're not dealing with climate change, or water, food, or all of these other pieces, it's not going to work—it's not really going to make durable change.'

Water governance provided the perfect lens for understanding these connections. 'Water is particularly interesting because it's a very overlooked resource. Growing up in Nevada, [I saw that] water is monetised, it is scarce, water rights are a big thing. Moving to British Columbia revealed different assumptions. You don't pay for water—it is just seen as a very endless resource.' Yet the apparent abundance masked deeper inequities: 'Indigenous communities in Canada don't have drinkable water in many places. Water is a really good indicator of how well our system is working. Are you valuing things properly? And are you valuing people properly? Does everyone have access?'

The transition from academic insight to professional commitment came through an energy-policy class. Though leaving university initially created what she describes as a directional crisis. 'I studied human geography, which is not really a well-known subject matter in North America,' she explains. 'It really boils down to the study of everything that's wrong with the world—it's very nihilistic in many ways. I graduated in a bit of a depression around like, "What's the point? We're all going to die anyway."'

Her first job at a climate and energy think tank provided crucial experience, where she started to see different pathways and why it was worthwhile 'to not let perfect be the enemy of the good.' This insight about the relationship between idealism and effectiveness would prove central to her approach to building Student Energy.

Student Energy had already existed for some time but had operated entirely through volunteers until receiving funding for staff in 2014. When they hired her in 2015 as community manager, she had no idea the executive director was planning on leaving. Almost overnight, she became executive director, inheriting 'six months of runway and a bit of enthusiasm to build on.' This experience taught her about what she calls 'founders and builders'—a distinction that has shaped her understanding of how sustainable change happens.

'It's a different personality type,' she explains. 'The former executive director really knew she was a founder, and potentially wasn't the right person to build. Founders are great, but they're often glorified in a certain way, or given expectations for skill sets they do not have. Founders excel at that initial spark to get things off the ground. They require a lot of dopamine, a lot of rewards, feeling that things progress super-fast. Building requires more sustained effort, more of a long game. It's the 10-year horizon versus the 2-year horizon.'

Her success in building Student Energy from a volunteer organisation to a global community reflected several factors, beginning with being 'young enough not to know that I shouldn't'—the advantage of naivety that allowed her to attempt things that more experienced leaders might have considered impossible. When living in Honduras, youth leadership programmes in Latin America provided distinctive preparation. Her high school experience in Carson City, Nevada, provided crucial education in diversity and privilege.

'I think it gave me a lot of exposure to the idea that there were many ways to live life and that there were many layers of privilege that I had. All of my early experience really added up to being able to build Student Energy in a more holistic way. And then I was also lucky that in my first jobs I had such inspiring superiors, from whom I learnt a lot about leadership and management. It helped me to adopt good guideposts on what good looks like, how to build an organisation, how to empower young people, and actually listen to what it is that they need, and build for that.'

The people who most shaped Meredith's understanding of leadership combined practical wisdom with willingness to challenge conventional thinking. Particularly influential was Juanita Sundberg, who taught Indigenous rights. 'She was just very bold, she would always kind of say what needed to be said. It was one of the first times that I'd really seen someone throw the rulebook aside and kind of charge forward with what they felt needed to be done.'

She learnt to distinguish between people who have 'shiny-penny syndrome'—constantly excited by new ideas but unable to implement them—and those who have 'really mobilised all kinds of different things.' One particularly powerful mentor was Katie Sullivan, whom Meredith met at her first speaking engagement with Student Energy. When she discovered that Sullivan had extensive United Nations Climate Conference of the Parties (COP) experience, she simply asked for help: 'I sat down next to her at a panel and learnt she had done a lot of COP work in the past. I was going to go to COP21 in Paris, but didn't know anything about COPs. So I simply said: "Could you explain to me how it works? I have no idea". You just have to admit you don't know what you're doing, and then people will often help.'

Meredith's analysis of urban sustainability reflects this practical orientation. She's witnessed debates that illustrate tensions between ideal solutions and practical realities, including in a current role on a local council committee: 'The town has a city planner who is convinced everyone should bike everywhere. But there's a lot of hills, there's a lot of rain, and there's tonnes of children. So how do you balance between the trade-offs that individuals are making, and designing for life that is more sustainable?'

Her current work in industrial cleantech represents an attempt to understand these challenges at a larger scale. Having spent years focused on community engagement, she became 'fascinated by all these really entrenched systems around things that we truly need, like steel, cement, and critical minerals.'

She continues, 'I believe the more exposure you get to other people, the more resilient our communities become, so when climate emergencies happen there is a kind of support network, a kind of faith in other humans. I think there are just core commonalities, things that almost everybody in the world cares about. Things that are very basic. Like wondering whether children have housing, whether they feel connected to something. I think every climate policy, every city should be able to drill down to those. You can build nuance from there.'

The challenge of creating such integration leads to her most trenchant critique of existing power structures. Working with governments, philanthropy, or large organisations, she feels the problem is not lack of good intentions but lack of experiential understanding of how communities work.

'Meeting people where they are is quite a stretch for most folks. I found this particularly true in centres of power. Even highly intelligent, well-educated leaders often struggle to recognise that everyone's life is not like their life, leading to well-intentioned interventions that miss the mark because they're designed around assumptions that don't match lived reality. It is actually valuing people just for being people. When that doesn't happen, it's very hard to get to solutions that work for everyone rather than just privileging those who already have access to resources and power.'

Her choice to remain in Squamish territory reflects commitment to 'working with what's in front of you and learning from that,' rather than constantly seeking perfect conditions. The town provides daily experience of sustainability principles, with every service within five to 10 minutes' walking distance, informing her broader thinking about how cities can support both individual flourishing and collective transformation.

Looking towards the future, Meredith's vision for effective change combines the community-building insights she developed through Student Energy with the systems thinking she's exploring in cleantech. But she also advocates for planned obsolescence. 'I think the ideal of being a charity is actually working yourself out of business,' she argues. 'Ideally Student Energy is no longer necessary at some point in time because young people have integrated access to the opportunities and resources they need'. This vision requires the courage to recognise when 'we did the thing' and when it is time for closure rather than allowing an initiative simply to continue to exist, challenging fundamental assumptions about institutional permanence and suggesting that the most successful organisations might be those that plan for their own eventual irrelevance.

3

Yunus Arikan (Türkiye/Germany): Dreaming Human-Scale Cities

Abstract Yunus Arikan is an environmental engineer from Middle East Technical University Ankara, Türkiye. Since 2009 at ICLEI – Local Governments for Sustainability, he has led global advocacy enhancing local government engagement in sustainability governance and has served as a UNFCCC Focal Point. His advocacy embraces stubborn optimism and the empowerment of younger generations.

Keywords City networks • Climate advocacy

2 February 2024, Online from Bonn, Germany

Pain still echoes through Yunus Arikan's (Fig. 3.1) voice as he describes the volleyball accident that has temporarily forced him into unexpected immobility. 'I have never screamed like that before. It was such a shock,' he recalls, his sprained ankle keeping him anchored to his Bonn home office. In this moment of forced stillness, Yunus finds himself contemplating rootedness—a recurring theme that has defined both his personal

C. Oke, J. Verbeeck, *City Leaders for Sustainable Futures*,
https://doi.org/10.1007/978-981-95-9544-0_3

Fig. 3.1 Yunus Arikan

journey and his vision for urban futures, a duality between tradition and transformation and the plains of Ankara and the green spaces of Bonn.

'I was a consultant at the time of my first Conference of the Parties (COP) [COP 9] in 2003, in Milan. There I heard about expert trainings by the Bonn United Nations Framework Convention on Climate Change (UNFCCC) secretariat. Eager, I just jumped on this new topic….' Although we are talking via Zoom, it is Rheinaue Park in Bonn to which he invites us. Here, 20 years ago, he first imagined a different future. 'During these 2004 UNFCCC training sessions, I was sitting on this lawn, thinking what would happen if I started to live in this city,' he recalls. Now, 15 years after making the German city his home, the park has become both origin story and touchstone, representing everything he believes cities can and should be: spaces where human connection flourishes, where the built environment serves rather than dominates natural systems, and where the pace of life matches the rhythm of genuine community. It reflects principles he would later advocate for in his work: accessible green space integrated into the urban fabric, for both solitude and community gathering (Fig. 3.2).

The memory of that moment shifted his perspective from seeing cities as problems to be solved through engineering interventions to understanding them as complex ecosystems requiring careful nurturing. His 15+ years as advocacy lead at ICLEI – Local Governments for Sustainablity, the global network of cities and regions committed to sustainability built on this experience.

Over more than three decades he had seen Ankara, his hometown, transform from a republican capital into what he describes as an increasingly challenging urban environment: 'Starting from '94, the city was in the hands of a fundamentalist conservative leadership. We simply lost faith. Life in the city, politics,…it all became really terrible. I was in the first two or three years of my marriage. In a couple of years, I would be probably walking my children to school, seeing them grow up…but it was not going to be in the Ankara I saw then.'

The transformation Yunus describes is universal. Many cities have witnessed how political shifts reshape not just governance but the very

Fig. 3.2 A picnic in a park, representing a memory of Rheinaue Park, Bonn, mentioned by Yunus Arikan

character of public spaces and by extension the possibilities for encounters across differences. The Ankara of his youth had been a place of intellectual ferment and cultural diversity, where Atatürk's republican vision had created space for experimentation and debate. The political changes of the 1990s brought what he experienced as a gradual constriction of possibilities, where public spaces became less welcoming to certain kinds of expression, where the rhythm of urban life increasingly reflected conservative social values. Beyond policy changes, it became a deeper transformation in urban culture, in the kinds of conversations possible in cafes and parks, and in the diversity of people who felt comfortable occupying public spaces.

> 'So at some point, instead of regular trips to Bonn, there was the point where I felt I would not return to Türkiye if there was no substantial change.'

In the early 1990s, whilst studying at Türkiye's Middle East Technical University, Yunus was already questioning prevailing urban paradigms through his involvement with the country's first eco-village initiative, Hocamköy. For that he and fellow students, driven by a vision of bridging traditional wisdom with modern innovation, ventured into the bare plains surrounding Ankara. 'We were the first to construct windmills and photovoltaics in villages,' he reflects with both pride and nostalgia, describing how 'we were trying to combine the "traditional" and the "modern", to create alternatives' that honoured rural knowledge whilst embracing technological possibilities. This early experiment contained the seeds of what would become his life's work: the pursuit of what he calls 'human-scale cities,' urban spaces that combine community bonds and connection to nature found in villages with the freedom and diversity of city life.

Working with rural communities, Yunus learnt about traditional building techniques that maintained comfortable indoor temperatures without mechanical heating or cooling, about agricultural practices that worked with, rather than against, natural cycles, about social arrangements that balanced individual autonomy with collective responsibility. These were not romantic returns to the past but creative adaptations of traditional knowledge to contemporary challenges.

It is a vision that stands in stark contrast to the megacities he believes have become unsustainable, leading to his conviction that 'I really believe a city should not be more than 1 million or so in order to be sustainable'—a number that emerges from lived experience of how size affects infrastructure and the very fabric of human relationships, the possibilities for genuine democracy, and the connection between urban dwellers and natural systems. It is not anti-urban sentiment but a carefully considered position based on decades of observing how cities function at different scales, how governance becomes more distant and bureaucratic as population increases, and how social problems become more intractable when communities grow beyond the scale where individual relationships can provide the foundation for collective problem-solving.

The story of his childhood neighbourhood in Ankara serves as a powerful illustration of urban transformations he has witnessed globally. He recalls living in a middle-class area that bordered what were then called 'gecekondu'—informal settlements that appeared overnight, where people were growing their own food. Yet the prevailing attitude among his neighbours was dismissive: 'They were starting to say, "Those people occupy our land, earning money and now mess up our city."'

Twenty years later, he witnessed those same plots transformed into what he calls 'gardens of stone'—with the city spending fortunes creating artificial versions of what was lost, building new developments 20 kilometres further out that attempted to recreate the very village-like qualities they had destroyed. This pattern demonstrates the tragic irony of urban development that destroys existing wisdom in the name of progress, only to spend vast resources trying to artificially recreate what has been lost, usually in forms that capture the aesthetic but miss the deeper ecological and social functions that made the original settlements sustainable.

His perspective on urban identity has evolved significantly during his years in Europe, shifting from seeing Türkiye as a bridge between East and West to embracing a more nuanced view centred on Mediterranean culture. 'Mediterranean includes Europe, Mediterranean includes Africa, Mediterranean includes Arabic communities and different religions,' he explains, 'and that's exactly what Europe has to be prepared for.' This understanding of cultural complexity informs his vision of cities as spaces of meeting and possibility, where difference becomes a source of strength rather than division. The Mediterranean framework offers an alternative to the 'clash of civilisations' narrative, suggesting instead possibilities for synthesis and mutual learning.

This cultural perspective has influenced his work on urban sustainability in profound ways, leading him to advocate for approaches that honour the diverse knowledge systems that migrants and refugees bring to European or other cities in the global North, rather than simply seeing them as populations to be integrated into existing urban systems. In his experience, some of the most innovative urban sustainability practices emerge from communities that maintain connections to rural knowledge whilst adapting to urban conditions, creating hybrid approaches that formal planning often fails to recognise or support.

He speaks with particular passion about the 'village institutes movement—Köy Enstitüleri' in early Republican Türkiye, which Yunus describes as bringing modern technology and cultural opportunities to rural areas without uprooting communities, demonstrating that modern technology could be adapted to the local tradition. 'The institutes were closed in the 1950s under pressure from conservative forces that saw them as too radical, providing a cautionary tale about the political challenges that alternative approaches to development often face.'

The conversation turns to memory and loss as Yunus reflects on returning to Ankara only once since leaving, for his 20th university anniversary in 2014. 'I was literally lost on that day. I just...I never thought I would be lost in Ankara because I thought I knew every corner of the city, but I simply couldn't find myself.' This disorientation in a once-familiar place he sees happens in so many of the world's cities, where urban history is erased in the name of progress and the landmarks of personal and collective identity disappear beneath waves of development that prioritise economic growth over cultural continuity.

Yet despite witnessing such losses, Yunus finds hope in younger generations and the explosive impact of young people like Greta Thunberg and her climate activism. He recalls seeing her at COP24 in Katowice in 2018, 'addressing the COP when there was no one in the room,' and then witnessing the transformation a year later in Madrid, when 'the room was so full when Greta was speaking they had to block the doors.' This rapid shift represents the kind of change he believes is possible when new voices emerge.

The conversation turns to his own children, revealing the personal stakes of his work. He speaks of initially hesitating to have children, not wanting to 'create a new victim' in a world facing climate crisis: 'I knew the 2020s, 2030s will be terrible for kids.' Yet life had other plans, and now his experience as a parent has deepened his commitment to urban transformation. 'They will change it because they don't have another option left,' he reflects, recognising that young people today possess a clarity about environmental challenges that previous generations lacked.

As we discuss maintaining hope in the face of climate crisis and urban degradation, Yunus emphasises the importance of what Christiana Figueres calls 'stubborn optimism'[1]—not blind faith but a committed stance towards possibility. He acknowledges the constraints that come with age and responsibility: mortgages, family obligations, and professional commitments that can make it harder to act boldly.

These insights have led him to focus on what he calls 'enabling work'—creating conditions where others, particularly young people, can take the kinds of risks and pursue the kinds of innovations that his own circumstances no longer permit. He frames it not as a limitation but as an opportunity to use his experience and networks in service of emerging leaders who bring fresh perspectives.

His vision for urban transformation extends beyond physical infrastructure to encompass new ways of thinking about economy and society, seeing potential in the emerging movement of wealthy individuals pledging significant portions of their resources to address global challenges. 'These are new innovations,' he observes. 'Again, there's so many routes for the revolution. It is not limited to only one.'

The term 'revolution' obviously should not be applied lightly, but Yunus wishes to acknowledge the gigantic scale of change introduced by urban and/or local and subnational actors into the global sustainability agenda over the past decades that he witnessed, shaped, and led through

[1] Figueres, C. (2024). Why a mind-set of stubborn optimism about the climate crisis is needed, now more than ever. *Bulletin of the Atomic Scientists*, *80*(1), 38–40. https://doi.org/10.1080/00963402.2023.2293576

his work at ICLEI as the Focal Point of the Local Governments and Municipal Authorities (LGMA) Constituency to the UNFCCC Secretariat. He remembers that whilst 'local authorities' were one of the nine Major Groups introduced to the global governance through 1992 Earth Summit, a discussion on the contributions and needs of local government, and eventually all types of subnational governments, to national commitments and global goals, was a dangerous taboo for many, especially in the climate space. Thanks to the coordinated, focused, constructive, and persistent efforts of the networks, this perception gradually but firmly changed, starting from the contested Copenhagen conference in 2009, formally cemented in the 2015 Paris Climate Agreement and expanded in its implementation phase. Yunus is confident that this new positive mindset will position multilevel collaboration and urbanisation, like two sides of the same coin, as a powerful lever in responding to the growing challenges in the second quarter of the twenty-first century.

As carnival music begins to drift through the streets of Bonn, Yunus recalls his own arrival in the city on carnival Sunday, confused by the masked revellers until he realised it was not just another day in his new home but a celebration of community and possibility. This memory leads him to reflect on the importance of cultural traditions in creating resilient urban communities, understanding how cities must remain spaces of encounter and celebration even as they face unprecedented challenges of climate change, inequality, and social fragmentation.

4

Claude Borna (Benin): The Village Builder

Abstract Claude Borna serves as Managing Director and Chief Innovation Officer of Sèmè City, Benin's flagship knowledge and innovation hub mandated by the government to drive human capital development. Drawing on experience at Deloitte, Amazon, and Sony Pictures, she leads practice-based education platforms aligned with employability and industrial needs.

Keywords Innovation • Knowledge economy • Capacity building • Entrepreneurship

6 February 2024, Online from London, England; Updated December 2025

London's vibrant character immediately comes to life through Claude Borna's screen as she dives into our conversation and describes the city she has spent so much time in (Fig. 4.1). 'I moved to London in 2000, actually. Wow. That's 24 years ago, and I was just blown away by how

C. Oke, J. Verbeeck, *City Leaders for Sustainable Futures*,
https://doi.org/10.1007/978-981-95-9544-0_4

Fig. 4.1 Claude Borna

people in London just mix. It's just a very eclectic and creative environment that I have always found to be extremely inspiring.' She carries with her the energy of someone whose identity spans continents yet whose heart remains anchored in the transformative work happening in Benin, a West African nation where the stakes of development are measured not in abstract policy frameworks but in the immediate possibilities available to young people seeking to build different futures. Claude was born in France, lived in the United States, went to university in Canada, and then moved to London before moving back to Benin.

Initial work in consulting left her frustrated. 'They teach you to give advice, but not do it, because you have to leave the client before they do whatever you're advising them to do. Even if you're advising them bullshit, you have to pretend this is the answer to all their problems, and then you can blame execution.' Later Claude worked at Amazon. In those early days, she experienced a different model based on the 'frugality leads to innovation' approach, where the door-desk that employees built themselves became a symbol of resourcefulness. 'When you have scarce resources, you use them a lot more wisely. And you actually learn to, well, innovate in a box … these constraints are exactly the things that make you be even more resourceful, even more able to push the boundaries.' Her later experience at Sony included witnessing 'the fall of a system' based on selling the same content multiple times. Her suggestion to explore African markets was dismissed as too risky, too poor, and not worth the investment—the kind of thinking that Claude's current work seeks to challenge.

When asked where she would meet us if not limited to a digital environment, Claude resists the premise entirely. 'I'm right here and now,' she states firmly. 'London offers that mix of creativity, diversity, wanting to be more sustainable and make more sense for people who live in London, which for me makes it a beacon of inspiration and innovation. When in Benin, there is a weight on my shoulders, which I don't have here. I think my creative alter ego is probably more inspired and unleashed here. Probably my answers would have been different, if we would have this talk being in Benin.' Benin offers her 'a feeling of belonging that is very, very distinct' combined with 'a responsibility that comes to me when I get here.' This requires being 'on your best behaviour' because 'it's a small country, full of small towns, where everybody knows each other, so people know you, they know your family.'

The accountability extends to professional obligation, where 'the [digital and creative innovation] work that I'm currently doing is serious work, and it's a lot of responsibility to ensure that we do it right.' The seriousness emerges from temporal urgency. 'We had 60 years of political unrest and people coming into power not doing what they should have done,' she reflects. Leading a knowledge and innovation hub, tasked by the current presidency, she wants to ensure that current efforts contribute to genuine transformation. Despite past mismanagement, community life has never died out. She switches the conversation to her sister's rooftop, 'where it's warm and where you can see the city, where you hear some Afrobeat music and you can be surrounded by the people you love'—using it as a metaphor for cultural continuity.

In the late 1990s, her sister started a clothing line using raffia—a natural fibre traditionally used in West Africa for everything from construction to wedding attire, creating 'a local take on the miniskirt, asking local raffia artisans to team up with sewing designers.' She continues, 'Now everyone's talking about eco-friendly fashion and circularity and sustainability in fashion…; those words didn't exist yet. But years later seeing people like Stella McCartney going and sourcing materials from Africa … isn't this what my sister was doing many years ago, I wondered?' Her sister continues this work today, 'looking at doing textiles from fruit fibre—banana, coconut … it's a technique that originates from the Philippines. So she went there, and learnt how they were doing it. And now they're trying to do it with a designer in Benin.' This South-South knowledge exchange represents what Claude sees as essential initiatives for African development (Fig. 4.2).

'For Benin, time is of the essence. Our young ones are having serious issues. So let's just do our best to not only just catch up on what we should have done in the past whatever 20, 30, 60 years, but leapfrog to the point where we're equipping them with what they need to be fit for the future. This is serious work. There's so many issues to be fixed, stuff that we can't take lightly. But it's not like having a lack of freedom. It's liberating to do that work. Because we had constraints before when basic stuff wasn't working, you know, water, electricity, infrastructure…education needing to be fixed.' At Benin's Sèmè City's Open Innovation Park, Claude and her colleagues made radical choices. 'We told those young ones at the very beginning, "This is your place, you run it. … We don't have much money to give you and you guys don't want to be bogged down by all the administrative procedures. So figure things out."' The approach evolved into recognising that 'had we wanted to do more engineering of what they're doing, I think we would have failed because at the end of the day, they use the resources that they have in a very interesting way and it limits waste—waste of everything. You know, waste of talent, waste of time, waste of material.'

Fig. 4.2 Illustration of sustainable clothing textile raffia as mentioned by Claude Borna

The young people created a skills-trading system where 'they started trading skills in terms of, okay, this is what I need, this is what you know, this is what I know and this is what you need. So why don't I teach you something and you give me something back?' This informal economy of knowledge exchange created learning opportunities that no curriculum could have designed.

She evokes the images of three young innovators. First, Reine 'went to med school, didn't finish because she didn't think it was for her and was, well, lost in a way.' Claude saw the misalignment between social expectations and personal calling. 'I think it reflects a situation that many of our young ones are in. They go on a path because that's what they're told they should do. She thought the right thing for her was to become a doctor. She's a very smart girl, but that wasn't the path for her. Our system doesn't provide human answers for that.' The Open Innovation Park provided what formal systems could not.

Mohamed's story represents an even more dramatic disruption. A graduate of Sèmè City's computer-science programme—'which is a three-year programme, and it's very different because they don't have professors, they have coaches. It's learning by project, it's peer learning and learning by doing'—he embodied problem-solving over credentialing. '[The programme] creates a profile of people who question everything. At the core of what they do is problem-solving.'

'Mohamed was good, you know, so I thought, "Let's hire him and let's put him to work at the Sèmè City open park," which wasn't even officially open at that point. We were, however, mostly looking for a manager, and I remember saying, "Hell no, he's just a graduate. He's 20!" I don't even know if he was 20 at the time. Without any experience running anything. So when he told me that he wanted to manage the place, my reaction was still: "Are you kidding me? Right?"'

The search for someone with appropriate experience lasted a year. Mohamed informally managed the space with such success that he forced Claude to confront her preconceptions. 'He just ran with it and did it so brilliantly that eventually he [formally] became manager. He told me, "Look at you with your preconceived ideas of you need x years of experience to manage the place. Blah Blah Blah."' More importantly, Mohamed's management style embodied unexpected values. 'He's the one who initiated the trading-skills thing. He just ran with it and, in doing so, gave the place its soul. It's a place of deep empathy now; they don't leave anyone behind. Last year, for example, they started working with kids who have special needs.' The team 'went through the whole summer with these kids and they did a fantastic team effort … without us giving them a manual.'

Abby rounds out the trio as 'an expert in mechanics and robotics' who 'joined by entering a hackathon' and now runs 'the mechanic lab over there.' Together these three young people represent the future of African innovation—'amazing, resourceful young ones who stop at nothing. You go to see them with a problem and they're like, "Don't worry, we'll fix it, we'll find a solution." I think we should give them more power.'

The generational analysis underlying Claude's optimism reflects her observation that young people who experienced COVID-19 during their formative years developed different relationships to responsibility. 'This generation…they're the ones who were hit hardest by COVID because it happened when they started university. So they didn't have the same university experience as we all did … the partying and the social things that you do, they didn't have that.' Rather than a deficit, Claude recognises this as creating different strengths: 'I find they have that sense of responsibility that I was talking about earlier that I'm not necessarily seeing with the millennials.'

Claude's approach—creating enabling conditions rather than imposing solutions—emerges from analysing why previous development efforts failed. 'What's happened before is that we thought we had all the solutions, and we tried to impose the solutions. I think we now realise that sometimes we need to take a step back and say, "Listen, we're enablers."' The role requires different humility: 'If you're in the public sector, yes, you have the opportunity to build infrastructure, to provide the basic things, but offering things like an internet connection and other basics that help people find solutions, this is what we should do. Giving them the means to find solutions that work for them.'

When asked who she would add to a superhero dinner party, in addition to her colleagues spoken of earlier, she mentions Michelle Obama for her empathy and architect Lesley Lokko for her intellectual rigour and cultural grounding. But most of all, she says, she would invite 'a bunch of 14- to 16-year-old Africans who are ready to transform the world, who are born without any limiting beliefs. Those who don't know what is often supposed to be impossible.'

The metaphor of the village that runs through Claude's thinking provides an alternative to individualistic and bureaucratic approaches, where 'in a village, it's an organised community and everyone has a role. Everyone knows why they're important in the village. And therefore, the whole system rests not just on one person, but on everybody. And everyone is linked.' This vision of villages as organised interdependence offers a framework for sustainable development that honours African social traditions whilst engaging with contemporary challenges.

5

Hans Bruyninckx (Belgium): The Quiet Revolutionary

Abstract Hans Bruyninckx is Professor of Global Environmental Governance at the University of Antwerp and former Executive Director of the European Environment Agency (2013–2023), leading the organisation during the European Green Deal's development.

Keywords Planetary boundaries • Biodiversity • Green infrastructure • Systems thinking • Environmental governance

14 February 2024, Online from Antwerp, Belgium

'If you live the life that we live and have careers that we have, with most of us living or working in cities, you are almost by definition connected to cities. Over the years I've spent a lot of time in cities and worked on cities and with cities around the globe. But to be honest, I'm not a big city person. I don't particularly like big cities. Just because of the noise. And I don't like big crowds either. So when abroad, I usually look for the

C. Oke, J. Verbeeck, *City Leaders for Sustainable Futures*,
https://doi.org/10.1007/978-981-95-9544-0_5

Fig. 5.1 Hans Bruyninckx

sort of third-tier cities to go to because they are more at human scale. But yeah, I mean, if we're going to change this society in any sustainable direction, it's going to happen in these large, expanding cities.'

Surrounded by the books that have shaped a lifetime of thinking about planetary boundaries and human possibility, Hans, Fig. 5.1, settles into our conversation from Mortsel, his own small town outside Antwerp, Belgium.

'My father was somebody who would read a lot, and he always took me to the annual book fair in Antwerp. In 1972, when I was eight, we came home with three books which, in hindsight, say something about many of the things I have done afterwards. We bought the *Times Atlas of the World*. That was my request. I wanted a big atlas. Then he also bought for me a book about threatened species, with nice pictures of African and Latin American animals. The third one was a translated version of the Club of Rome report. If you take those three together, you almost have what I built a large part of my career on. … I didn't read the Club of Rome report then of course, but my father did and talked about it for a long time. Being a kid in the fourth, fifth, sixth grade we would always watch the news at a fixed time—you couldn't organise the evening

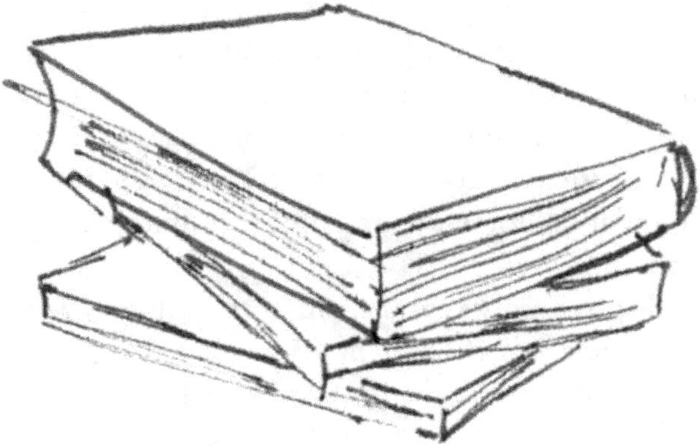

Fig. 5.2 A stack of books, illustrative of their importance to Hans Bruyninckx's story

rhythm in our family without [that]—and so I would sit next to him and he would explain the news to me.'

He recalls how childhood influences have perhaps proven more predictive than formal career planning. The atlas led to endless games mapping global trade routes and calculating gross domestic product (GDP) per capita for different countries (the name of the game was Magellan), providing early education in the interconnections between geography, economics, and politics that would later inform his understanding of how environmental challenges cross national boundaries and require international cooperation. The threatened-species book provided early exposure to extinction dynamics that would later inform his work on biodiversity policy, whilst the Club of Rome report represented something deeper yet: an introduction to systems thinking that would become central to his professional identity. After his father passed away, he inherited more books. A selection of them is on one of the shelves behind his desk (Fig. 5.2).

The natural experiences that complemented his formal education provided different but equally important lessons, where a teenage Interrail trip with a friend took an unconventional route. Rather than heading south to beaches and parties, like most young Europeans, they decided to see 'how far north can we get by train.' The journey above the Arctic Circle in Scandinavia, hiking with inadequate equipment through reindeer territory, provided what Hans describes as an encounter with 'overwhelming nature' that complemented his intellectual understanding of environmental systems with a life-changing appreciation for natural complexity and beauty. This experience of being dwarfed by natural systems whilst simultaneously feeling deeply connected to them provided an emotional foundation for environmental commitment that academic study alone could not supply.

Hans would spend a career translating between the abstract language of scientific research and the urgent realities of policymaking, and he remains fundamentally grounded in the conviction that knowledge carries responsibility. After a decade of living in Copenhagen, where he led the European Environment Agency during some of the most critical years for the development and implementation of environmental policy, he currently resides in Belgium again. Looking back, he has seen the scaffolding and implementation of the European Green Deal that represents the world's most ambitious attempt to achieve carbon neutrality whilst maintaining economic prosperity and social cohesion. Hans, however, speaks with self-deprecation about his role as merely 'a drop in the bucket.' He was never much of a spotlight seeker, but when given an audience he would often speak with a certain impatience, even a polite radicalism.

He invites us for another walk along memory lane, to Fort Collins, Colorado, at the foot of the Rocky Mountains, where he lived for four years in the mid-1990s as a PhD student and instructor. 'Of all the places I have lived, it's the place where I feel most at home. Academically speaking, Fort Collins for me opened up the world, after coming from the conservative Belgian academic environment, where as a young researcher you had the feeling that you were light years below the professors. For my PhD studies within the field of international relations, I wanted to do research on the link between environmental degradation and violent conflicts, but at that time that proposal was not taken seriously. So rather than in Leuven, I ended up doing my PhD in the US. Fort Collins was then about 85,000 people. It's now about 170,000. It has grown rapidly, but it's still a city that is connected to its environment, to the great outdoors. There, I was taken seriously. Me, a young foreigner with an accent, teaching students on global politics and even on American government.

It didn't take much time for professors to become colleagues and friends. I still really love that place. Whenever I go back to Colorado and get off the plane—it's the only real way to get there—the first thought that comes is "I'm home."'

The academic culture stood in stark contrast to the rigid hierarchies of traditional European universities, where advancement depended more on connections and narrowly defined credentials. The Colorado experience would determine his subsequent career. He moved from there to his first teaching job at a small liberal arts school in upstate New York, where his task was not only to convey knowledge but also to help create citizenship. But it was not the unfolding research and professional pathways that would shape many of his later steps but family life.

'In a span of 17 months, me and my wife had our first two kids, and they were both mentally handicapped. So we decided to come back to Europe. We wanted to go home. Also, I didn't want our kids to depend on charity. I wanted to move back to a society where there is a proper social system. And so we came back to Belgium.' He would take many of his insights back to Europe. His work at the Leuven University and Wageningen University focused on governance and institution building in complex policy settings: how democratic institutions can be reformed to address long-term challenges that transcend electoral cycles and national boundaries, how technocratic expertise can serve rather than replace democratic decision-making, and how international cooperation can be built through patient institutional development.

Herman Deleeck, his political science professor at Leuven, embodied everything Hans believes about the relationship between knowledge and social responsibility. 'He was an absolutely passionate teacher, able to convey that passion. But he also conveyed the message that science is fun. That science is societally relevant, not just a hobby thing, and that scientists have a responsibility in society,' Hans explains.

The classroom moment that crystallised this understanding has become legendary. Midway through a semester course with 120 students, Deleeck noticed restlessness and made an unexpected offer: 'Okay, I get it. You're getting tired listening to me. If you think you can do better, where my finger stops next week, you explain the next chapter in the book.' Without waiting, Hans volunteered. 'My fellow students thought I was completely nuts. "Are you serious?" Professor Deleeck asked. And I said, "Oh, I'm really serious." So the week after, I spent a full week diving into the material and going to the library and stuff but I did do the teaching. And I never stopped teaching or speaking in public ever since.'

Hans is refreshingly honest: 'I didn't have the working life that I had because I am so brilliant. I think you have the life you have because you are lucky, you work hard, and you're not risk-averse. And I'm passionate about things. You know, passion has been my fuel. But there are too many people in academia who think that they got there because they've got the best idea and the best concept and the best theoretical contribution. I think those are really, really rare people. Most of us have just been a bit lucky. I've bumped into not just one but several people who believed in me at the right moment and stimulated me.'

'When I came back from the US, with my PhD in international environmental politics, there was not a single course in Flanders taught on the subject, not a single one. So it took me 2 years until the Dean of the Faculty agreed to start a course on that. And for 10 years, I think I was the only one in Flanders teaching that course. There were years that I had 250 master's students in that course. They came from everywhere, all different faculties. Almost until I moved on to my job at the EU level [in 2013], some colleagues in political science still didn't accept that it was a serious topic.'

This relative isolation had unexpected benefits, as it forced Hans to seek international research funding and collaborate with scholars worldwide. He explains, 'I got a lot of international and European research money, so I carved out relative independence' as a co-founder of the Leuven Centre for Global Governance Studies.

The transition to policy implementation came through his appointment to lead the European Environment Agency during the crucial period when the European Green Deal was developed. 'To be able to do so, I was looking for peers: people who have really deep scientific knowledge but have the intelligence, the personality, the skills, to be connectors,' he says.

This emphasis on connection reflects Hans's fundamental critique of traditional academic culture. 'Value-free science for me is exactly what it says it is. It's science without value,' he argues, insisting that whilst scientific methods must remain rigorous, the choice of research questions inevitably reflects the researcher's values. 'The awareness that something needs to change if we don't want to hit the planetary boundaries is out there, but in a very crowded field of snippets of information, which is increasingly preventing us from thinking in big frames. Scientifically speaking, those big frames are stronger than before. Systems science is stronger than ever, and at the same time, we are in such a fragmented society when it comes to information and to attention that it's almost in opposition to each other. That is really concerning.'

When asked about the parallel with the noise he had mentioned at the start of our talk, he said, 'My personal approach is, I'm on no social media, so I don't waste any time on it. And I've decided that from the start; I didn't see any added value for me in it. So that is at least part of the fragmentation of time that I'm not part of. But in the end, it's about something much bigger. The current continuous flow of information is not only undermining our attention span but also prevents us from really discussing the bigger issues. Because every five minutes there is something else that draws your attention, which in the end is really rather insignificant…I don't know. I have recently become a board member of an organisation that is working on democratic innovation. One of the questions there is whether going to a voting booth every 4 years really is democratic or not. I think that this is a very limiting interpretation of what a healthy democracy should be. I'm really concerned about the quality of the democratic debate. That's also why I give quite a lot of public lectures also to young people and to broader audiences, not to specialists, with clear images, compelling storylines, and scientific information that speaks to people's values—because climate change is not a problem of physics, it's a problem of society. I try to engage people in a story about sustainability, which is at the same time a story about democracy and the European Union—as a social scientist, that's the agency that I have.'

Hans looks at cities as places that might offer healthy tension between creative disruption and collective responsibility. 'Cities are places where people meet and where you have creativity and also where you have—I think—the right amount of counterculture and anarchy, and that's I think what we need if you want social change. I do think that we look at cities too much through the lens of problems that have to be solved and don't give enough space to the creative forces we have in cities. At the same time, I also strongly believe that we cannot be blind to all the social issues that play in cities and that erode our social fabric. In the ideal world, you should not need a lock on your bike. It's these small crimes, for which there are no consequences, that are undermining the public and the social cohesion. It creates mistrust, it leads to stigmatisation, it leads to individualism when it comes to property and the public space. I think it is fundamentally weakening our societies.'

Hans's responses illustrate broader tensions facing professionals committed to global environmental challenges whilst remaining grounded in particular communities. 'Life happens, and it throws all sorts of stuff on your plate that is unexpected and that you're not responsible for—or only partially responsible for,' he reflects. The organisation his family founded to provide housing and community for adults with mental disabilities represents Hans's practical commitment to the values he promotes professionally. The net-zero carbon house they built for their children and nine other adults represents the kind of innovation he believes cities need more of: practical demonstrations that sustainable living can enhance, rather than restrict, quality of life.

6

Paola Flores Carvajal (Bolivia): Weaving Indigenous Wisdom into Urban Futures

Abstract Paola Flores Carvajal is a Bolivian industrial engineer. As Head of Product at DeltaX, she uses AI to digitise logistics. A WTO Young Trade Leader, she advocates for Indigenous recognition, climate justice, and inclusion of women and youth in global trade.

Keywords Climate justice • Social sustainability

25 January 2024, Online from Sucre, Bolivia; Updated December 2025

Slightly under the weather from a recent cold caught during travels between cities, Paola Flores nonetheless radiates unmistakable energy as she transports us 20 years into the past to a place where her journey as an Indigenous climate activist first took root (Fig. 6.1). It has been a while since we last spoke, and her welcoming presence gently engulfs us again. She has that quality of someone who carries traditional wisdom whilst navigating the cutting edge of technology. Someone whose deepest

C. Oke, J. Verbeeck, *City Leaders for Sustainable Futures*,
https://doi.org/10.1007/978-981-95-9544-0_6

Fig. 6.1 Paola Flores Carvajal

insights emerge from understanding how childhood experiences in rural Bolivia can inform responses to global environmental challenges that affect billions of people worldwide. Her story unfolds as a bridge between worlds: her grandmother's village house and the conference halls of international climate and trade negotiations, between Quechua language traditions and English-language policy documents, and between the patient rhythms of traditional Indigenous Bolivian stories and practices and the urgent pace of contemporary climate activism.

If Paola could choose our meeting place, she would build a bridge back to her five-year-old self, to a rural Bolivian district where mud-brick houses rise from the earth itself—structures built from the ancient marriage of soil and water, a building technique passed down through generations that embodies everything she has learnt about sustainability, resilience, and the integration of human needs with natural cycles. In this memory space, her grandmother sits in the yard of their home, a steady presence against a backdrop of mountains, with a river flowing in the near distance.

This imagery for her represents the importance of understanding that Indigenous knowledge requires active preservation and revitalisation, that linguistic diversity connects directly to ecological diversity, and that the traditional languages that often accompany traditional ecological knowledge could inform responses to contemporary environmental challenges. Even at five years old, Paola felt stirrings of purpose that would guide her entire life, observing the basic needs and challenges that her community and others in Bolivia faced and storing these observations like fuel for future action.

'This is the place, the community where I was born but also where a certain wondering emerged. I love this place because it taught me a lot of things. But what was the reality of Indigenous communities? How did their traditions and rituals allow them to sustain everyday life? At some point, I just knew that I wanted to do something, I wanted to change something, because I felt that there were more things outside of my everyday reality that people in the communities of Bolivia should know and should be empowered to do in the future. I didn't know what exactly, but I was sure that many things needed to be better for us to have a better quality of life.'

The object Paola chooses to represent the beginning of her sustainability journey carries profound personal significance: a small conference identification badge, worn and treasured, bearing her name and the title 'Magnifica Warmi.' This represents her first climate project working with Indigenous women, launched with just $500 in funding and embodying the resourcefulness that would become characteristic of her approach to development work that prioritises community engagement over expensive infrastructure. 'Every material you're seeing here is recycled,' she explains, holding up the carefully preserved badge. 'Every person on my team had this personal identification. We made everything by hand.' (Fig. 6.2).

Fig. 6.2 Illustration of the Magnifica Warmi conference lanyard mentioned by Paola Flores Carvajal

The project emerged from collaboration with her best friends, all young women, who wanted to create an almost instant, tangible impact in their communities. This was after an earlier attempt at a more technical project—building machines to recycle plastic bottles—proved too time-consuming for their desire to see immediate results that could benefit the women they were working with. 'It was really special, because it was my first time specifically talking about women in the climate space,' she reflects. At 19 years old, Paola and her team ventured into unknown territory, working with women two or three decades their senior, not in theoretical exercises in climate education but in intimate exchanges where traditional knowledge met contemporary environmental challenges, where the wisdom of experience encountered the energy of youth activism.

The project revealed something crucial about the intersection of Indigenous wisdom and climate action that would inform all her subsequent work. 'The mothers knew what was happening in their communities, but they didn't know how to express it,' Paola explains. 'Climate change scientific theories were too complex for those who lived the facts.' Simultaneously, and not surprisingly, the young activists discovered that their role wasn't to teach but to learn—about traditions, beliefs, and the profound role of women, especially mothers, in environmental stewardship that extends across generations. This early experience exemplified the power of shared meals and traditional hospitality. When the mothers prepared food for the community, following ancient customs, Paola witnessed how climate action could be woven into existing cultural practices rather than imposed as an external agenda. 'It is a common tradition in Bolivia that one person cooks for others. You eat the food, and you don't leave anything on your plate,' she explains, describing customs that her grandmother had practised with neighbours, sharing food on special days as an expression of community solidarity that created bonds extending far beyond individual families.

The experience awakened deep pride in her Indigenous identity that would sustain her through subsequent challenges. 'I have to say that I feel really proud of being of Indigenous descent. When I saw the traditions, I told myself, "This is my place, this is where I belong,"' she shares, describing belonging as 'a call in your blood, something awakening in your mind and offering your smile because you belong to a place'.

Her journey to this point was not without significant challenges. Growing up, Paola felt excluded in some spaces for her Indigenous features. Yet rather than diminishing her connection to her Indigenous identity, her experiences strengthened her resolve. The experiences taught her that 'no matter how difficult the situations were, it really gave me a lot of strength to say, "Come on, we can demonstrate many things." Those challenging experiences gave me the strength to be here right now and confront the problems that I had in those years.' A transformative moment came at age 15 when Paola attended a summer camp at the University of San Diego in California. The experience opened her eyes to renewable energy possibilities. It crystallised her childhood sense of wanting to 'change the world' into a concrete understanding of how science and technology could serve that purpose whilst also showing how 'diversity was a way to do things more powerfully.' More recently, Paola has ventured into the technology sector, facing new challenges as an Indigenous woman in a male-dominated field.

Currently working as a Head of Product at a Bolivian technology start-up whilst maintaining her climate activism, she embodies the very intersection she seeks to create: where Indigenous wisdom meets technological innovation, where traditional knowledge informs contemporary solutions, and where ancient understanding of natural cycles can guide modern approaches to sustainable development. Three days before our conversation, Bolivia celebrated its renaming from the Republic of Bolivia to the Plurinational State of Bolivia, in recognition of the country's diverse Indigenous communities, which comprise over 50% of the population. We speak about this official acknowledgement of diversity existing alongside ongoing struggles over resource extraction and environmental protection, creating contradictions between constitutional recognition and economic practice. The contradiction is not lost on Paola. 'How does a nation honour its Indigenous heritage whilst simultaneously threatening the lands and lives of Indigenous environmental defenders?' she wonders. Her vision for addressing these challenges emerges in her description of the team she would assemble to rebuild society, bringing together 'someone from the past but also someone from the future.' The Indigenous person from the past would carry ancient wisdom that contemporary Indigenous communities have lost through modernisation pressures, although Paola notes, 'They have more wisdom than we have, even right now.' The person from the future would be 'an engineer, a scientist' with a political background, someone who understands that 'there's no way in our actual society to make changes if not with policies,' bridging the gap between innovation and implementation that characterises so many failed sustainability initiatives. Crucially, she emphasises the importance of including her friends: young people working in the climate space who are 'really talented' and 'doing incredible things,' where her insistence on friendship as foundation for effective climate action reflects understanding that personal relationships and emotional connections drive the kind of sustained commitment that technical solutions alone cannot generate.

Her message to young people reflects hard-won wisdom about the importance of courage and persistence: 'Young people are needed and need to be treated as equal partners in decision-making. We have been living the problems.' This intergenerational perspective brings urgency to climate action whilst also carrying forward traditional knowledge that older generations possess, creating bridges between ancient wisdom and contemporary innovation that could inform more effective approaches to sustainable development.

The power of intergenerational knowledge transfer becomes visceral when she describes her grandmother's influence, years after her passing. 'She gave me her cultural heritage. She gave me her blood. She gave me her traditions. In Indigenous communities, you will see no discrimination—no matter where you are from. Once you get into a community, you are welcome, you are loved in some way.' This principle of welcome and inclusion stands in stark contrast to her experiences in urban spaces, suggesting different models for city development. 'That's something that cities need to learn—how to embrace the diversity.'

Paola maintains hope grounded in understanding that change emerges from individual action aggregated across communities and generations. 'If you see a problem, just solve it,' she concludes with a simplicity that nevertheless is underlaid by a profound complexity. This directive emerges from lived experience of discrimination and exclusion, from witnessing environmental destruction and community displacement, and from understanding that effective solutions must bridge traditional wisdom and contemporary innovation. Through Paola's eyes, we see how cities might become not just sustainable but regenerative, drawing strength from Indigenous wisdom whilst embracing technological innovation, creating spaces where the dreams of five-year-old children in rural Bolivia can shape the urban futures that billions of people will inhabit in the decades to come.

7

Winston Chow (Singapore): Life Along the Riverside

Abstract Winston Chow is Lee Kong Chian Professor of Urban Climate at Singapore Management University and Co-Chair of the IPCC's Working Group II (Impacts, Adaptation, and Vulnerability) for the 7th Assessment Cycle, which includes leading the development of the IPCC Special Report on Climate Change and Cities. His research focuses on urban climate risk, vulnerability, and heat island adaptation. He advises international urban resilience projects and teaches climate change and urban sustainability.

Keyword Climate risk • Adaptation strategies • Urban resilience

22 December 2025, Online from Singapore

Winston Chow is fast to answer the question of where we would meet, if not online. It would not be in gleaming conference halls or academic corridors (Fig. 7.1). Instead, we would sit along the banks of Bangkok's Chao Phraya River near Saphan Taksin BTS station, at a simple table

C. Oke, J. Verbeeck, *City Leaders for Sustainable Futures*,
https://doi.org/10.1007/978-981-95-9544-0_7

Fig. 7.1 Winston Chow

beneath an ordinary fan with cold beers sweating in the tropical heat. The river flows as metaphor through the entire conversation—from mountains through agricultural lands to dense urban centres and finally the sea, connecting places and people, carrying both life and waste, subject to seasonal floods affecting millions. Understanding Bangkok's climate vulnerability requires understanding this river system, how upstream use affects downstream flooding, and how communities have adapted to living with water's rhythms.

'Bangkok is my favourite city in the entire world,' Winston declares. Not Singapore, where he's built his career as one of Asia's leading urban climate scientists, but Thailand's capital—with all its contradictions, chaos, the food and fashion but also 'warts and all, the seediness, the nightlife.' To him, Bangkok represents 'the ultimate representation of what a city should be, a place where pockets of sustainability coexist with urban challenges, where community organisations rally for liveability alongside the density that characterises rapidly growing Asian cities.'

Listening to Winston paint Bangkok city life, one feels how he is not drawn to sanitised urbanism but to cities in their full complexity. Notice his words: 'should be,' not 'could be.' Bangkok is not an aspiration but a reminder of what cities actually are: spaces where sustainability must engage with real human desires, not just technical solutions. Winston's connection to Bangkok runs deeper than cuisine or culture. He's conducted research there for 20 years, brought students, and built relationships with universities. 'With every trip I also brought some clothes back. The shirt I'm wearing actually came from JJ Market, from Chatuchak,' he notes with his typical grin.

But beneath this familiarity lies a deeper education about urbanisation itself. Winston is a child of Southeast Asia, watching cities transform across generations. In his younger days, his family journeyed from Singapore to his father's ancestral *kampung* (village) about 45 km northwest of Kuala Lumpur in Malaysia. 'Over the years, that small settlement, which used to be an eight-hour trip by car, became a seven-hour one, then six hours as things became more urbanised. The north-south highway connecting Kuala Lumpur to Singapore was developed. The *kampung* changed from a small collection of buildings to a town centre that suddenly expanded.'

This was his first classroom in urban transformation, his first textbooks these annual pilgrimages, which revealed development's accelerating pace. 'I could see with my own eyes how the nature-scapes I was used to seeing as a kid disappeared and became concretised and asphaltised. But I've also seen the benefits—electrification, communities partaking in development.' This dual awareness—seeing both environmental costs and human benefits—would stick. It sensitised him to 'the trade-offs of development and sustainability, which fed into broader theories and frameworks of urban resilience and urban studies.'

Probably just as foundational, Winston's team members insisted that he take time to actually experience the cities where he attended climate meetings—Sofia, Riga, São Paulo—taking in urban life with a personal gaze, not just a professional one. Walking became both respite and research. 'While walking, I gauge how walkable a city actually is.' He acknowledges his privilege as a six-foot-three, 200-pound man who does not generally feel threatened, but getting lost in the urban maze allows him to read a city's character.

'I'm constantly looking out for things—odd things sometimes. Being from Singapore, where everything's organised underground, I notice how power infrastructure is sometimes draped across cables, how drainage is done, but also what the overall vibe of a place is. How rituals like Oktoberfest transform Munich every year.' In São Paulo, he pondered how the city maintained trees that sometimes completely overgrew power and communications cables. His reflections show how his urban wanderings are never purely recreational and how his work is not merely professional. Winston is always reading the urban landscape for climate vulnerabilities and resilience features. This blending of perspectives reflects how he resists the compartmentalisation that treats climate science as purely technical, separated from lived experience. The best urban climate science, he shares, emerges from sustained engagement with places as they actually are, not how models represent them.

The seeds for intertwined approaches on sustainability were planted early. At age 11 or 12, he discovered science fiction and graphic novels. 'I was consuming anything that was space opera, anything with ecosystems that clearly are collapsing.' But one particular comic book, *2000 AD*, which depicted a dystopian future, catalysed his environmental awakening.

'That comic book series had a massive impression on me. Because of the artwork—the drab Brit-Cit grey, the massive scale and squalid landscape of Mega-City One—but also the messaging: that actions we do or don't do will have ramifications for urban living in the future.' Years later, working on climate communication, Winston would remember how powerful images can convey urgency in ways that data sometimes cannot.

Images, and the possibility of change they make tangible, can stick. For his object of inspiration, he shares a generic image of an aerosol can as a symbol of what is possible and talks of the hugely successful international environmental campaign to ban chlorofluorocarbons to save the ozone layer (Fig. 7.2). He also speaks about a water-saving campaign in Singapore that left a mark in the early 2000s 'the so-called "NEWater." Back then and still now, Singapore's survival is so dependent on water

Fig. 7.2 An aerosol can, mentioned by Winston Chow as a symbol of inspiration

resource management. After finishing my university studies as an undergrad, I was fascinated by the deep-tunnel sewer systems, the engineering process of filtration, and the UV treatment to make the wastewater more sterile than distilled water. It was one of the breakthroughs in terms of Singapore's long-term water resource planning, of making sure we didn't have to import water anymore from Malaysia.'

'NEWater is basically recycled sewer water. But it wasn't the technological development that I was interested in. It was in how Singaporean governmental agencies messaged and communicated the acceptance of new water to the general public. Back then, it was called "shit water" because people were scared—"I am not drinking somebody else's piss"—until the then-Prime Minister during the equivalent of the State of the Union address just took a swig of it and said, "It's perfectly safe."'

When we talk about societal tipping points, sometimes the messenger is more important than the message in trying to effect behavioural change and to make sure that certain aspects of resilience can materialise, especially when magnified as good practices by cities or, in the case of Singapore, city-states.

This early exposure taught him to think in feedback loops and cascading consequences, to understand environmental problems as systemic. Yet he found himself at the beginning of his professional career, trained as a pure scientist, focusing on quantitative data and urban climate modelling. It took an additional nudge from early mentors during his PhD at Arizona State University to embrace complexity. 'You need to start looking into the social, economic, demographic, political chaos that all happen in that same system. And part of the fun of interdisciplinary work is looking into how methods for both physical sciences and the social sciences can be combined together to give you new insights. It taught me not to fall into the trap of thinking of a city as a homogenous entity.'

Winston's work with the IPCC—contributing to multiple assessment reports and now the Special Report on Climate Change and Cities—represents boundary crossing between science and policy and research and application. His particular expertise in urban heat vulnerability has taken on increasing urgency as cities face more-frequent and more-intense heat waves. In that respect Singapore shows both opportunities and challenges. As a wealthy city-state with near-universal air conditioning, it differs dramatically from other Southeast Asian cities where heat vulnerability intersects with poverty and outdoor labour. This comparative perspective makes Winston acutely aware of how climate impacts are mediated by social conditions. 'The same heat wave will have very different consequences depending on whether you're in an air-conditioned office or working outdoors, whether you have access to cool refuges or live in informal housing with metal roofs.' Technical interventions like cooling centres or green infrastructure must be designed with specific populations and their constraints in mind.

Field experiences during his time in Phoenix, Arizona, told him not to overlook the role of governments. 'The way to reach out to vulnerable communities is via the leaders or the representatives of these communities. It's often more effective to go into churches or community centres, communicating the incoming 115–120 degree Fahrenheit weather, but also telling them: "You can have this shaded safe space open for people to come in during summer. We can have people walking around who can provide water, who can provide shelter, or a place of air conditioning, so on and so forth." You won't be able to do this sort of very effective resilience, applied resilience work, just by ignoring culture. If so, you're not going to get successful policies, full stop. We still underestimate the importance of messages coming from somebody who is viewed to be local or speaks the local language. So culture does play an important role in that.'

On whether cities can help bring global stability, Winston's reply is both cautious and frank: 'Everything's fractured and focused on nationalism, which regrettably, we're seeing right now. Powerful nation-states enable this, with just a few individuals basically dictating stuff or being the figurehead of a particular political vision or paradigm. So where would cities play on this global stage? Can they? Can they be the sort of white blood cells to push back? Will that happen as a sort of culture of cities? The amount of people that live in cities outnumber the people who don't live in cities? But it will be about more than one or two "city" visions, because there's always numerous and separate regional and cultural issues that come to the fore.' He adds, 'We're always asking for the city's voice at the table. The problem is, which voice and which city has that seat at the table. I struggle to think if today there is a viable alternative paradigm of an urban network, urban culture, urban response. It's difficult to see that, but you never know.'

Returning to science fiction and future ecosystem collapse, we wonder about resilience and how to prevent or recover from collapse. Winston's answer divulges his strategic approach to collaboration. His first choice surprises us: 'Elon Musk.'

He acknowledges the controversy and then explains: 'It's always good to pick people with contrary views who think they are smarter than you. It's oftentimes useful to persuade, engage, and also to not lose sight of the simple fact that in the end, Elon Musk puts on his pants the same way I do. I would like to know if there is a way to get through and confront them with their own ego—if you're so smart, figure out a way to solve this mess. Sometimes for these type-A personalities, you throw them a problem, for them to take as a challenge to solve.'

For strategic thinking, he chooses chess masters: Garry Kasparov particularly, for how he's navigated politics alongside chess. For governance, Lee Kuan Yew, Singapore's founding Prime Minister. Then comes a crucial shift: 'Lastly, to make sure it's not a sausage party, I need grounding.

I don't want just all males and old people. So I would also have Marie Curie. As I tell my kids, don't listen to old men telling you what you can or can't do. Curie didn't. She won two Nobel prizes, one in physics and one in chemistry, to a large extent based on intellectual curiosity. She can teach us about determination combined with intelligence.'

He then pauses, his voice becoming more soft: 'But also…somebody from the future. It has to be either my kids or somebody else's kids—maybe I can be generous and say my as-yet-unborn grandchildren. I don't want to look towards the past. I need to look to the future, having them as ballast to make sure that what we're doing is important, is critical.' This shift—from provocateurs and strategists to future generations—explains the heart of Winston's approach. All the strategic thinking, all the engagement with power ultimately serves one purpose: creating cities where his grandchildren might flourish. The as-yet-unborn become moral anchors, reminding us that every infrastructure investment creates the world that future generations will inherit.

Winston's career displays a pragmatism born from working in the messy space between knowledge and action. Just as he emphasised giving oneself 'time to enjoy being in a city, in a new place,' he returns to the necessity of personal connection to place as a foundation for professional effectiveness. Climate work, for him, is fundamentally collaborative. No single discipline has all the needed knowledge. Effective interventions emerge from combining technical expertise with local knowledge, scientific projections with community priorities. Sometimes this means engaging with people you fundamentally disagree with, learning from those who operate according to different values. 'The need for difficult conversations—sometimes people tend to not place a premium on that.'

He knows that perfect solutions do not exist, that trade-offs must be negotiated, and that incremental progress often matters more than comprehensive transformation. Cities are never static—always becoming, always negotiating between past forms and future possibilities. Winston has learnt to work within existing systems while looking for opportunities to shift them, to build relationships that make future cooperation possible. And always, the as-yet-unborn grandchildren hover in the background—moral compass, future consequence, the reason all this difficult work matters. Sometimes, that work is best approached over cold beers beside a river, in conversation with people who care deeply about places and are willing to do the patient, difficult work of helping them thrive in an uncertain future—always with one eye on present challenges and another on the as-yet-unborn grandchildren who will inherit the cities we create today.

8

Nathalie Flores (Dominican Republic): Forward from Moca

Abstract Nathalie Flores works at the intersection of climate policy, carbon markets, and financial services. As Vice President—Carbon Markets at StoneX Group Inc., she supports scaling carbon mechanisms across compliance and voluntary frameworks. She previously served as Director of Climate Change for the Dominican Republic and was elected Vice Chair of the UNFCCC Subsidiary Body for Scientific and Technological Advice.

Keywords Climate justice • Multilateral governance

28 January 2024, Online from the Dominican Republic; Updated in 2025

Surrounded by the organised chaos of international moves, Nathalie Flores settles into our conversation sitting between packing boxes, her infectious energy undimmed by the practical demands of relocation (Fig. 8.1). At the same time, she never loses sight of Moca, her tiny

C. Oke, J. Verbeeck, *City Leaders for Sustainable Futures*,
https://doi.org/10.1007/978-981-95-9544-0_8

Fig. 8.1 Nathalie Flores

hometown in Espaillat province, where her journey towards international climate leadership began with parents whose love provided the only foundation needed to pursue dreams that seemed impossible from such a remote starting point.

She would not take us to the Dominican Republic if she could choose our meeting place. London holds the crown—'It's like an upgraded New York City,' she states without filters, asserting that it creates perfect conditions for the kind of cross-cultural dialogue that defines her work as Vice Chair of the UN Climate Change Subsidiary Body for Scientific and Technological Advice. In this imagined London meeting, we would find ourselves at Sketch, her favourite afternoon tea venue, surrounded by an environment 'completely pink' (though recently changed to yellow), sharing 'scones and all these little sandwiches and tea' whilst ballerinas bring champagne and conversation flows about 'how I feel about the world and sustainability in my whole life and career.'

The often light stories and inviting winks reveal something essential about Nathalie's approach to diplomacy: the recognition that the most important negotiations often happen in informal settings, that cultural exchange and relationship-building matter as much as technical expertise, and that creating spaces for genuine human connection enables the kind of trust required for effective multilateral cooperation. New York occupies a different space in her imagination—'a city that shaped me,' she reflects, calling it both home and not home, 'a city of dreams I love with all my heart—but a closed space, not a city that I want to go back and live in again.' Metropolitan centres provide essential opportunities and perspectives but remain places of passage rather than permanent belonging. Singapore rounds out her urban trinity, representing something closer to utopian possibility: a place that 'has the weather of the Dominican Republic, the development and the "future look" in sustainability terms that I don't think any country in the world has and also the British system around roads, language, and everything like that.' This vision of Singapore as perfectly integrated urban sustainability provides a template for what she hopes other island nations might achieve.

She has barely finished depicting Singapore when she holds up a memento from the 15th anniversary of the Obama Foundation: a simple plaque bearing the word 'Forward.' Despite almost everything being in boxes, she has not forgotten our request. 'I don't want to politicise too much,' she explains, 'but this specific memento, and the fact that I saw it again while I was packing, inspired me a lot to this conversation.'

> 'My country is very, very dependent on the US. With the largest Dominican immigrant community established in New York and New Jersey, we even have a little Dominican Republic in Manhattan. What I wanted to be in life was a diplomat. I was never too involved in issues related to the environment or sustainability. I wanted to do international affairs, I wanted to travel the world, I wanted to represent my country.'

The word 'Forward' embodies this determination to move beyond the limitations of circumstance, to create opportunities where none existed, and to represent not just personal ambition but collective possibility for young people from small places who dare to dream of global impact. The

Fig. 8.2 An illustration of the Forward poster Nathalie Flores provides as her object of inspiration

journey from that small town to global diplomacy required what she describes as pure determination supported by parental love (Fig. 8.2).

'I'm from a small town called Moca. My father is an agricultural engineer, my mother an educator, and the only thing me and my sisters had was their love to try to pursue our dreams, because we had a lot of limitations in terms of where to go to school, what to study. They did everything to ensure that we have a future.' Her parents provided both practical understanding of environmental challenges and deep commitment to learning as transformation, creating a household where intellectual curiosity was nurtured even when resources were limited.

A decisive moment came during a high school field trip to the Dominican Congress in 2003–2004, organised by her civics professor. The experience created complete infatuation: 'I completely fell in love with the idea of becoming a congresswoman one day.'

Initially, this political awakening led her towards mass communications, but an introductory course showed immediate misalignment: 'I didn't like mass communication. So without paying attention to my parents, without paying attention to my professor—I decided, "I have to go to law school."' The switch required negotiating with parents concerned about their eldest daughter leaving home at 16, leading to the compromise that she would study law as foundation for specialising in international affairs and diplomacy.

'In the end I never liked law school either,' she admits with disarming honesty, 'but it was my foundation platform to be a diplomat and to defend my country.' The transformation came in the semester before graduation, when she enrolled in environmental law as an alternative to disappointing international law courses. 'Finally, something I liked from law school! I ended up doing my bachelor's dissertation on—what is now not such a good analysis—the application of environmental agreements in the Dominican Republic. So I think we see how things went from high school, until the moment environmental law and environmental international agreements made me fall in love with what I do now. This is what I lived for.'

This academic awakening coincided with a practical opportunity when she secured a position as assistant to the first climate change director at the Ministry of Environment and Natural Resources, where the ministry—being the national focal point and thus responsible for all the environmental agreements—provided what she describes as 'the perfect ecosystem for somebody that wants to implement, to regulate, and also defend the country's interest at the international level.' It combined local implementation with global negotiation in ways that matched her hybrid interests perfectly.

Her civics professor planted the original seed of public service, whilst her environmental law professor raised awareness that environmental law was a reality.

Perhaps most influential was Patricia Abreu, former Vice Minister of International Cooperation, who at 21 became Nathalie's boss. 'She was super young, but she was always encouraging us to apply to all these international cooperation trainings from Japan to Brazil, to explore South-South cooperation, sending us to places across the globe. "Train yourself," she would say. "Go ahead." Only 23–24–25 years old, but she would send us to our first Conference of the Parties (COP), telling us we would not be going to negotiate as we were still being trained but that we could learn a lot by just taking care of the Dominican Republic Pavilion. It's important also to understand we all have a role to play in this.'

Even booth management at that first COP—COP16 in Cancún, Mexico—provided essential education in how international negotiations actually function, demonstrating how seemingly mundane responsibilities can become gateways to understanding complex systems. In the years following, working with Federica Bietta and Kevin Conrad from the Coalition for Rainforest Nations in New York helped her adopt 'other perspectives from international academia, bankers, and the private sector.' These experiences taught practical skills but also provided models for how to maintain integrity whilst operating across different institutional contexts.

Her role as Vice Chair of the UN Climate Change Subsidiary Body for Scientific and Technological Advice represents what she calls 'the cherry on top of my diplomatic dreams,' combining her original diplomatic aspirations with deep environmental expertise developed over 13 years of patient engagement with both technical and political dimensions of climate action. It showed her both the possibilities and limitations.

> 'I always like to provide people with the context of the United Nations being a club of governments to understand that we cannot control everything, even if we want to. The United Nations is powered by the governments and for the governments. I think a lot of people do not understand that or do not acknowledge that.'

What helps her avoid despair is seeing that change does happen. 'Of course, from a government point of view, you sometimes feel trapped needing to go by the many rules or because they stick to known patterns. But yes, when it comes to the climate topic, I feel we are still taking baby steps. Thirty years is nothing for the magnitude of the problem. We are still very far away from where we want to be, where we need to be. Every time citizens want change, they need to change themselves as well. You cannot have a lot of one ideology or way of operating and at the same time a lot from another.'

Her favourite example of successful international environmental cooperation is the Montreal Protocol, which she calls 'one of the few environmental treaties that actually work when it comes to implementation.' The protocol's success in eliminating ozone-depleting substances provides a template for climate action: 'If we got to work to eliminate the gases that were affecting the ozone layer, I'm sure we can do the same for greenhouse gas emissions as well, if we apply the change and have the willingness to do that.'

She insists on focusing on breakthrough achievements. 'COP28 was not a failure, because this is the first time ever that we actually have text on fossil fuels. You are not going to tell all the fossil fuel countries that they cannot use petrol anymore, and you are going to use electric cars right away, and you are not changing everything overnight and with one decision—that's just not going to happen. I think the problem is that people put a lot of hope in the Conference of the Parties. And it goes beyond the Conference of the Parties. Take, for example, the United States. Even if the federal government decides to pull out of a specific treaty, then there is still the state level, which often is still building capacity, working with universities, doing research, or even putting together programmes like the California cap and trade.'

> 'In practice, the Conference of the Parties is a group of ministers that decide on the final text. There's just too much expectation about something that is not legally binding. What is going to have an impact is what you do in national laws and structures to actually regulate and honour international agreements. Just taxpayer money is not going to resolve the climate crisis; you need the private sector on board to do that.'

Her approach to communication reflects this understanding. 'I think there are two pillars that we need to actually make people understand what sustainability is: raising awareness first, building capacity second. People see the environment as a form of activism, because they often cannot see what the concept of sustainability is at its core. If we do a survey in the Dominican Republic, there's no Dominican that can relate tourism with the environment. Currently in our country, the speech is that tourism is one thing and protecting the environment something completely different. But well, how are we having tourism if we don't protect the environment? How are we going to have a healthy coastal marine ecosystem for you to come to the Dominican Republic if we don't have regulation and protect the environment? So, there is a need for a strong campaign of shaping a sustainability culture.'

The power of education becomes visceral when she describes her young relatives. 'Valentina [who was eight years old when this interview was recorded], when she was four, she told me that I shouldn't drink Coca-Cola because of the plastic bottle,' she recalls with obvious pride.

When children grow up understanding sustainability as normal rather than exceptional, when they question adult practices that harm the environment—such cultural shifts create the foundation for policy changes that might otherwise remain impossible.

Her use of social media reflects strategic commitment to positive messaging. 'I'm not going to share anything that is negative in my social media. I want to raise awareness but from a positive perspective. When we get to the negative side of the issue, then you are already putting your reader in a sense that they need to come to you to argue.'

She wraps up by referring to Gabriela Márquez, who from early 2025 has been the Climate Change Director of the Dominican Republic. In 2012, when Nathalie had only one year experience in climate work, 19-year-old Gabriela wrote via social media: 'I've seen you on Facebook and Twitter, and I want to be like you—my dream is to be like you.' Years later, Gabriela ended up working as Climate Change Director with Nathalie, bringing complementary skills that balance Nathalie's encouragement-focused leadership with more analytical, systematic approaches.

Nathalie wraps up her passionate pledge with exactly that: passion. 'You cannot work if you don't have passion for any specific topic. It doesn't matter if it's selling hot dogs in New York or cleaning the beach in Panama.' Without passionate commitment, 'the job becomes monotony,' but with passion, even the most technical international negotiations become opportunities for meaningful impact that can transform both individual lives and collective possibilities for addressing shared challenges.

9

Cynthia Houniuhi (Solomon Islands): The Land Remembers

Abstract Cynthia Houniuhi, Solomon Islands climate justice lawyer and former President of Pacific Islands Students Fighting Climate Change, who spent her youth walking land boundaries with elders in her matrilineal community, orientated her legal strategy when 27 University of the South Pacific students petitioned for an International Court of Justice (ICJ) advisory opinion on climate obligations. In 2024, she testified at the ICJ, translating sacred Indigenous knowledge into international law.

Keywords Climate justice • Intergenerational equity • Climate litigation

29 December 2025, Online from Solomon Islands

Cynthia Houniuhi is calling in from the Solomon Islands, part of Melanesia, to the north-east of Australia (Fig. 9.1). If she could choose where we would meet, it would be beneath a banyan tree—an *abalolo* in

C. Oke, J. Verbeeck, *City Leaders for Sustainable Futures*,
https://doi.org/10.1007/978-981-95-9544-0_9

Fig. 9.1 Cynthia Houniuhi

her Malaita language—on the University of the South Pacific campus in Vanuatu (Fig. 9.2). The 83 islands that together form Vanuatu lie in the remote reaches of the South Pacific Ocean. The small developing island state is home to around 300,000 people and has put climate on the agenda of global and international courts. Under current scenarios, with average temperatures increasing by more than 2° Celsius, the Pacific island states are predicted to lose a fifth of their GDP annually due to climate disasters. Vanuatu is a leading voice for climate justice that, in an effort to achieve international accountability and the meeting of legal obligations, had the UN General Assembly ask the International Court of Justice (ICJ) for an Advisory Opinion on climate change obligations, seeking binding legal standards and clearer state duties to protect the climate system and address harm.

Fig. 9.2 An Illustration of a tree and bench, Cynthia's place of solace and inspiration

A green bench sits under its canopy, facing the lagoon. 'Usually when I'm stressed or just want to reflect, I go sit under this tree,' she explains. 'Here, I always think about the beauty of the world. Will generations that come after see the same things? When we go out there to talk about where we're from and what we see happen around us, people don't even know where Vanuatu or the Solomon Islands are. And so what we do is tell the story that the world is an island. Everyone is on that island, and though we are on the front line, we're all in this together. It's not about blaming and shaming. That was why we picked up this global and legal discourse, as an inclusive process, to offer people a real, tangible pathway that can be supported. That's why I would want us to do the interview there—to keep us grounded on the bigger picture.'

This orientation towards future generations defines everything Cynthia does. As founding member and former president of Pacific Islands Students Fighting Climate Change, she helped bring climate justice to the International Court of Justice. But the story begins in the hills of the Solomon Islands, with a child walking land boundaries that she did not yet understand she was being taught to protect.

'Most of the land in Solomon Islands is owned customarily by Indigenous people,' Cynthia explains. 'I come from a matrilineal system: the land passes through the women.' When elders show land boundaries, they always bring a young person. She was that child, walking without understanding why. 'When I grew up, I realised the long-term vision. We don't have written records—we pass down knowledge through stories. My being there meant when elders pass away, I am a witness.' Even in land disputes, you have a witness who will grow up and tell these stories. 'That moment stuck with me—we have a traditional system in place that is still future-proof.' She is, however, quick to add, 'My generation is losing that connection. Yet, we do not have to go far to learn about systems important for our people. We can learn them from our elders.'

Around fires on bamboo benches, while grandmothers cooked, Cynthia learnt 'custom stories'—origin narratives about how her people relate to the world. Her favourite is the one about sharks, the totem for the place where she comes from: 'We call them our guardians because we believe they are the two sons of the first woman who lived on the island, turned into sharks.' These stories teach seeing nature not as something separate but as a form of relationship. 'Not just to my family but to the entire island.'

She makes a crucial distinction: 'It's not my story, it's our story.' This collective orientation creates accountability. 'The media wants a hero, one person to speak, but we have this understanding that whoever speaks represents the collective. When we go back to our village, we will be held accountable by the whole island. So we do it with respect, care, and hope.' Cynthia offers an image capturing this ethic: 'When you go fishing, you pull five fish. When you come back, one fish belongs to you, the four others to your neighbours. Even though you did all the work, the rest belongs to others. It's just the way of living.'

That sense of collectiveness is best reflected in and by the oral traditions and the commonalities and universal truths they often address. 'A lot of people that have issues—for example, fighting for their land in completely different circumstances, not even climate related—they can understand what it's like to be determined, to fight, to be stubborn. And so continuing to speak to people and learning about their challenges, we often end up seeing similarities but also having just heart-to-heart conversations. And although people don't always insist on the storytelling, sometimes more on just sitting down and truly meeting, that's maybe how the most impact is made.'

Her time at the University of New South Wales in Sydney in 2023 challenged these norms, making her appreciate them differently. 'The more I go out into the world and learn more, the more I'm gravitating back. To move forward, I need to learn the systems already in place, because that's how we take ownership.' She pauses: 'When you step on the ground [with] bare feet—that's that connection right there. That grounding is important. Conflict is a common occurrence, and when it happens, the chief will give advice, which then brings harmony to the whole village. You could consider the ICJ the world's chief. We realised with Pacific Islands Students Fighting Climate Change, we're not on the same page in terms of international law, so we needed that clarity. So we journeyed as an international community to the equivalent chief of the world to give us advice, as we believe that it will give that harmony and start processes that we need and not incremental but ones that provide a baseline that's more certain than what we have at the moment.' When Pacific Islands Students Fighting Climate Change collected testimonials for the ICJ, Cynthia travelled to Areare in South Malaita, where chiefs hold sacred traditional knowledge. They shared it for the court. What struck her was that these chiefs see intergenerational equity—being fair to the next generation—not as an aspiration but as a 'sacred duty.' Climate change makes this duty almost impossible. 'You sever that duty when it's so hard to look after the land, when you have sea-level rise, frequent rainfalls, all the challenges.' Yet the duty remains. This shapes her message: 'Think about your grandchildren. When you leave this world, will you leave it with certainty, knowing you left it at a better place for them? It's about that care and love. That's the same thing we're doing, there's, of course, our island, but in the bigger picture we see the whole world as a community.'

Cynthia's path to law came from observing systems that do not reflect Pacific realities. The ICJ campaign became an exercise in reimagining how international law could work. Her analogy, discussed above, is powerful: in her village, when there's conflict, people go to the chief for advice. 'The whole logic is the same—the ICJ is the world's chief.' But translation poses profound challenges. 'How do you explain Indigenous knowledge in English to fit into recognised UN language? When the elders passed on old sacred knowledge for us to translate, I couldn't sleep at night. I was worried we would not do justice to it.'

Even the term 'climate change' does not exist in her language. 'Our people use words to help describe what's happening. They do not say "climate change."' When collecting testimonials, Cynthia and her colleagues avoided technical terms: 'We just wanted to hear what they observed before and what they observe now.'

When asked who inspired her work, Cynthia points to parents who taught service. 'My family is filled with nurses. I took a different road and went to law school, but it was seeing my parents serving the most rural, remote parts of the island.' Her father had a saying: 'Greatness comes from serving people.' This reframes the climate justice campaign: 'What is my contribution, my service? We're just part of the ecosystem. What is our part?' When asked how she holds the tension between responsibility and overwhelming reality, her voice becomes more serious. 'It's not easy. There are days when you just want to give up.' But connection sustains her—to land, community, and identity. 'What other choice do we have? We do not accept the alternative. Hope is a powerful belief.'

At a recent conference, a panellist asked everyone to breathe deeply and then said: 'That's it. That's what we should value. Once it leaves our body, everything else is just not worth it.' This grounds Cynthia when overwhelmed and sharpens her warning: 'People have pity and say, "Oh, the Pacific Islands, we are at the front line." In perhaps five, 10 or 20 years, other countries might be on the front line, and we do not wish this for you.'

When asked about her dream team that could help translate the sense of urgency into action, Cynthia's face lights up. 'We'll need a mansion!' She wants everyone who walked alongside her in the ICJ journey—Pacific Islands Students Fighting Climate Change, World Youth for Climate Justice from Europe, Asia, Africa, Caribbean, the teachers, lawyers, chiefs, young people, 'even young children with their placards.' She continues, 'We really need that reunion. We haven't had time to celebrate.' Her ideal gathering would be about acknowledgement. 'The Pacific way is to look people in the eye and say thank you.'

In panel debates she is often asked what she would do, if she had all the money in the world: 'I would bring all the people I need to say thank you to, to the Solomons, and say thank you in person. That's the best way to show respect—it was not just us, we built on the work of many others.' This collective orientation keeps returning as our conversation continues: the students who chose hope instead of despair. The elders who entrusted sacred knowledge. The parents serving remote communities. The chiefs maintaining intergenerational duty despite impossible circumstances. The young women standing between logging companies and ancestral land.

The ICJ opinion represents a beginning, not an ending. The work continues: teaching young people to maintain connection to land and language, working with women to reclaim traditional authority as landholders, and building Indigenous-informed systems. When Cynthia speaks of hope, she points to young people. 'I went into teaching because we have that opportunity to shape that thinking in a way that nurtures connection. They will be the ones—I look to them as hope.' But she's clear eyed: 'Influential people in the country are losing sight of these values. That's why we're working with young people, to bring that back.' Still, the foundation remains: the land, the custom stories, the sacred duty to future generations, the collective orientation that sees five fish as one for you and four for neighbours. These are not relics but resources for the future, systems that have proven resilient across generations and might yet prove resilient through the climate crisis—if young people can learn them, honour them, and adapt them.

Cynthia closes her eyes and sees the green bench under the *abalolo* tree, the lagoon beyond. Will those who come after see the same? The question is not rhetorical but operational, driving every decision about what to teach, what to protect, what to fight for, and whom to serve. She refers again to her father's saying: 'Greatness comes from serving people.' 'And so for this whole journey,' she says, 'it's about what is my contribution, what is my service? So if it's president of PISFCC, then so be it, if it's assistant lecturer, then that's it. It can be daunting, as you get a lot of criticism, even [from] the ones that are closest and do not understand. But we see them as opportunities, and so having that value of service really keeps you going. And not just ourselves but the wider ecosystem we are part of. I'm sure if we look in terms of science, you will see how the overall ecosystem maintains its own balance. We have our part to play as well.'

The question remains whether oral traditions and knowledge are passed on fast enough. 'The value that is emerging now is the recognition of not only elders as wisdom holders but also young people as change makers, as credible people. It is something that will shift our trajectory in terms of where we go in the next five or 10 years.' Apart from the intergenerational knowledge, it is also the land that remembers. And those who walk it have a choice: to be witnesses for generations who come after or to be the generation that broke the chain.

10

Sophie Howe (Wales): Guardian of Tomorrow

Abstract Between 2016 and 2023, Sophie Howe served as the world's first Future Generations Commissioner for Wales, described by The Guardian as 'the World's First Minister for the Unborn.' She held the Welsh Government accountable on how decisions affect future generations, influencing the United Nations' first Declaration on Future Generations in September 2024. Her approach emerged from Oxwich Beach gatherings where collective care shaped her understanding of intergenerational justice.

Keywords Intergenerational equity • Urban resilience • Collaborative governance • Wellbeing

29 January 2024, Online from Cardiff, Wales

Between the familiar chaos of the school run and settling into her home office, Sophie Howe embodies the intersection of intimate family rhythms and the broader work of public policy that she has helped reshape during

C. Oke, J. Verbeeck, *City Leaders for Sustainable Futures*,
https://doi.org/10.1007/978-981-95-9544-0_10

Fig. 10.1 Sophie Howe

her groundbreaking seven-year tenure as the world's first Future Generations Commissioner (Fig. 10.1). Having just dropped off one of her five children, she appears on screen with the focus and determination of someone who has spent years advocating for a different tempo in political life—one that considers the impact of today's decisions on children not yet born, on communities not yet formed, on challenges not yet fully understood.

The personal and political dimensions of her work intertwine in ways that reflect the fundamental challenge of future generations' thinking: how to maintain focus on long-term outcomes whilst meeting the immediate needs of constituents who face daily pressures of housing costs, employment uncertainty, and climate anxiety. Her own experience as a mother navigating these tensions provides both motivation and practical insight into how policy frameworks must be designed to serve not just abstract future generations but the actual children whose futures depend on decisions being made today.

We are meeting online, but if Sophie could choose our meeting place, she would transport us far from Cardiff's urban bustle to Oxwich, a pristine stretch of coastline in the Gower Peninsula near Swansea where she spent formative summers that would shape her understanding of both nature and human possibility. 'It doesn't look so different now as when it did when I was that eight- or nine-year-old kid roaming the sea and the cliffs. We used to go every weekend to this beach, just hanging out in nature and having long lunches that would go into dinners with friends from the local communities. There were probably about between 20 and 30 of us, a number of different families who would always be there.' She laughs at herself and at the rose-tinted glasses she has put on while sharing all of this, but the scene she paints is vivid: children roaming freely between sea and cliffs, their feet finding purchase on slippery rocks as they hunt for crabs in tidal pools, their voices carried on salt-tinged breezes, though she laughs, saying, 'I'm sure that with the British weather we often have that I've kind of made the images better than they were, because in my mind it were always long hot summers there.'

Oxwich is not mere nostalgia. For an only child who 'hated' that solitude, these gatherings represented a community that transcended individual family units, shaping her understanding of how sustainable communities function when they operate according to different values to those that dominate contemporary political and economic life. The families who gathered created temporary villages where children belonged to everyone, where safety emerged from collective care rather than individual vigilance, where the simplicity of shared meals and informal supervision enabled forms of freedom that seem increasingly impossible in contemporary urban environments. 'It is those unpretentious, cost-free days. There's something about them, being back to basics. The moment you realise: we don't need all this stuff to be happy.' She describes how barbecues on the beach and shared meals created bonds that would last long after the sand was shaken from swimming costumes, relationships formed through shared attention to place and to each other's children.

The contrast with her later experience growing up in Ely, a district of Cardiff that appears regularly 'in the headlines for the wrong reasons' with high unemployment, teenage pregnancy, and low life expectancy, provides a stark illustration of how geography and economics intersect to create vastly different childhood experiences within the same city. Living now in what she calls Cardiff's 'leafy suburbs, where everybody wants to live,' she sharply sees the culminating effects of discriminations against those living at intersectionalities they hardly ever escape. This juxtaposition between the preserved natural beauty of the Gower and the challenges of urban deprivation would become fundamental to her understanding of environmental and social justice, creating an intuitive grasp that disadvantages compound themselves—that living in poverty means not just having less money but breathing dirtier air, having fewer green spaces, and experiencing higher stress levels, all of which accumulate across generations in ways that simple policy interventions struggle to address.

The spark that would ignite her passion for global justice came from an unexpected source: the cultural phenomenon of Band Aid in the 1980s. As an 11-year-old watching Bob Geldof galvanise global attention around famine and poverty, Sophie glimpsed the power of popular culture to drive social change through what she describes as the first time 'that I and possibly large swathes of people from across the world were really thinking about poverty inequity and what was happening in other countries.' This moment of collective awakening, mediated through music and celebrity, taught her early that change often comes through unexpected channels and that the tools and potential for transformation extend far beyond traditional political processes.

The small Band Aid badge she still keeps embodies her understanding that cultural change often precedes political change and that shifts in consciousness happen through popular culture and peer networks rather than through formal education or government campaigns. This insight has informed her approach to implementing the Future Generations Act, where she recognises that legal frameworks alone cannot create the cultural transformation necessary for genuinely sustainable development (Fig. 10.2).

She recalls a telling moment when her teenage son had an epiphany about climate change not through her years of work as Future Generations Commissioner, but through watching *Don't Look Up* on Netflix. It led to

Fig. 10.2 An illustration of the Band Aid, Feed the World campaign of 1980s

a profound conversation 'as you usually do have with teens…at bizarre moments, when we were making breakfast.' 'It is absolutely awful,' she remembers him saying. 'They are not doing anything.' The irony wasn't lost on her: here was a young person who had grown up surrounded by climate policy discussions, yet it took a satirical film to crystallise his understanding of institutional denial and systemic failure. 'You've got to find a way of making people care, whatever that means,' she says, 'and there are various ways that you need to consider how you do that. Especially and increasingly, for our, you know, not even just our younger generations, I think we're all having a reduced sort of concentration span. So for some people, it's data and evidence and thousand-page reports; for others, it's the David Attenborough documentaries or just snippets of information or perhaps creative interventions.'

Her own path from being the first in her family to attend university to becoming the world's first Future Generations Commissioner reflects both personal determination and a vision for systemic change that developed over decades of practical experience with how public policy actually affects real communities. Early in her career as a local councillor, elected at the remarkably young age of 21, she encountered what she calls the 'utter madness' of public service delivery: systems designed around institutional convenience rather than human need. When dealing with issues around 'youth annoyance or antisocial behaviour,' Sophie understood that 'the kids were hanging around the streets because they were bored,' but the response from public services was to close youth facilities during school holidays when young people had the most free time and greatest need for structured activities. These frustrations with short-term thinking and siloed approaches would fuel what became her life's work: transforming governance to serve future generations through the Welsh Future Generations Act, which requires public bodies to consider the long-term impact of their decisions across social, economic, environmental, and cultural dimensions.

Creating new legislation proved easier than transforming embedded institutional cultures. The Future Generations Act came into being in 2015, with the recognition that legal requirements alone cannot shift institutional ethos that have been shaped by decades of short-term thinking and departmental isolation. The Act was initially completely misunderstood. Sophie describes the beauty of the Act as its function as a four-legged chair, where social, economic, environmental, and cultural well-being, as well as the intersections among them, need to be considered. The work requires both formal frameworks and informal relationship building, both policy mandates and cultural interventions that help people imagine different ways of thinking about their professional responsibilities. 'So anything but a tick-box exercise, rather a fundamental reorientation of how decisions get made. I can't count the times I have witnessed departments demonstrating compliance with the legal requirements whilst missing entirely the deeper intention behind the legislation. "Where is the impact we seek?", I have often exclaimed.' This isn't bureaucratic perfectionism but recognition that social, economic, environmental, and cultural well-being are interconnected in ways that resist compartmentalisation. Over time many departments did shift from considering isolated impacts to engaging in systemic thinking; this represents the kind of cultural change that the Act seeks to embed across Welsh public life.

'None of this is quick,' notes Sophie. 'It will be a generation in Wales, 20 to 25 years, before we can say that every decision is taken absolutely through this lens.' This timeline challenges every assumption about political progress, requiring leaders to plant trees whose shade they will never enjoy, to invest political capital in changes that may not show results until long after they have left office. The complexity of this approach becomes evident in current struggles over transport policy in Wales, where Sophie speaks with evident concern about a Welsh Transport Minister who recently took stress leave due to public backlash against sustainable transport initiatives. 'People don't want to have to drive at 20 miles an hour, and they do want their road to be built, and they're fed up with not having good public transport,' she acknowledges, illustrating one of the hardest aspects of long-term thinking: sometimes things must get worse before they get better, creating friction, discomfort, and resistance during transition periods between old systems and new ones. Stopping road building does not immediately improve public transport; reducing speed limits does not immediately change travel behaviour; investing in renewable energy does not immediately reduce energy costs. These intermediate steps can feel punitive rather than progressive, especially for people already struggling with daily challenges.

Yet Sophie finds hope in unexpected places, particularly in the changes she's witnessed between her eldest child's education and her youngest's, where her nine-year-old daughter participates in school uniform swap shops to reduce waste, lobbies against plastic milk cartons, and engages in food-growing projects—practical expressions of sustainability thinking that were absent from her older children's education just a few years earlier. 'I think there's something around joining the dots between all of these things and working out where the magic is. Where's the magic between them and how can we bring them all together?' The magic she refers to isn't supernatural but systemic: the emergence of new ways of thinking and being that arise when different approaches begin to reinforce each other, when creative interventions in healthcare settings help patients process difficult experiences whilst providing feedback to improve services, when school projects on sustainability create conversations that extend into family life and community practices, and when local food growing connects environmental awareness with health outcomes and community building.

Through her experiences raising her five children whilst advancing systemic change, Sophie embodies the intersection of personal and political transformation, where family life provides both laboratory and testing ground for the principles she advocates professionally. The breakfast table conversations with teenagers all become data points in understanding how cultural change happens at the most intimate scales and how individual behaviour change and institutional transformation must work together if sustainable development is to become more than policy aspiration. Her children's different responses to sustainability messaging, their varying levels of environmental anxiety, and their distinct ways of engaging with global challenges all provide insight into how future generations' thinking must account for individual differences whilst creating collective frameworks for action.

11

Indy Johar (UK): Weaving New Patterns of Possibility

Abstract Indy Johar is an architect, social entrepreneur, and co-founder of Dark Matter Labs, a field laboratory building institutional infrastructures for radical civic transformation in cities and regions. While he was studying at the University of Bath, a book about labyrinths and pyramids pivoted his understanding of transparency and asymmetric power and of intimacy and freedom.

Keywords Boundary spanning • Regenerative economics • Adaptive governance • Systems thinking

25 January 2024, Online from London, England

Appearing on screen not as the acclaimed architect and system innovator many know him to be but as a questioner of questions themselves, Indy Johar immediately challenges the very premise of our conversation when asked about what the ideal setting would be (Fig. 11.1). 'I really struggle with that question,' he says, 'because I don't like to romanticise the past.

C. Oke, J. Verbeeck, *City Leaders for Sustainable Futures*,
https://doi.org/10.1007/978-981-95-9544-0_11

Fig. 11.1 Indy Johar

I think it is one of the biggest traps we fall into.' This answer to what is, after all, a non-question, the immediate intellectual pivot from place to memory to the dangers of romanticisation, probably portrays Indy better than whatever any question and answer could have achieved. It already hints at how Indy moves through the world: less interested in answers than in the spaces between them. It will be a red thread throughout the conversation, the intricate patterns of possibility that emerge when we resist the urge to simplify complex realities into comfortable narratives.

The crystalline light of a winter morning in London might seduce one's thoughts to drift off, but Indy is that type of person who really demands full and open presence. His thinking seems to hopscotch simultaneously across multiple domains of knowledge and practice, weaving connections between philosophy and economics, between ancient wisdom and contemporary innovation, and between the immediate challenges of urban development and the vast scales of systemic transformation that our current moment demands. Conversations with Indy can sometimes be a roller coaster. Love it or hate it, not something in between. Yet at the same time, he can demonstrate a curiosity that is both a provocation and an invitation. It emerges from decades spent working at the intersection of architecture, technology, and social innovation, where his journey has taken him from designing buildings to designing systems, from solving spatial problems to addressing the deeper patterns of organisation that shape how societies function. Dark Matter Labs, the name of the organisation he stewards, could not have been better chosen.

London became his laboratory for understanding how complex systems adapt and evolve, where his work with Dark Matter Labs explores questions that traditional disciplines struggle to address: How do we design for emergence rather than control? How do we create conditions for collective intelligence to flourish? How do we move beyond the extractive logics that have shaped both our cities and our economic systems towards regenerative approaches that serve life rather than destroying it?

But if forced to imagine a possible setting for our conversation, it would not be London. Indy would transport us not to any existing city but to a place that exists only in the intersection of multiple possible futures, describing a vision that marries 'the walled city of Kowloon in terms of its sort of mass unleashing of different contributions' with '*Star Trek* operating in a different economy' and 'radical democracy.'

'Why Kowloon? I don't like to romanticise the past. I think we sometimes look for some perfect adult that happened in the past, which might be the biggest trap to fall into. So I really struggle with that particular frame of question to find an ideal conversation spot. I can obviously say Rome, where I did my master's programme, because you can swim through so much time. But it feels like a trap. What we're looking for is something else. What I'm more excited about is where we are, as the capacity to create society has been more democratised than it has ever been. And I don't mean the vote, I mean the actual means of construction of society. The distributed agency to construct the world around us, is, I think, quite phenomenal. Whether it's through service, through music, through buildings, through collective decisions, through different forms of buying, through volunteering. We have this plurality of making the city that we ourselves are part of, and I think we're living in a world where our means of societal contributions are so multiple … yet I don't think our cities are able to handle those means yet.'

'I would want to do this interview there, where the bustle of life and contribution is so rich, so integrated and woven … and where the ecologies of transaction are operating with different currencies and different relationships of care. Where we are living in hybrid richness of human and non-human systems or entangled machine systems being eloquently entangled in a beautiful kind of tangled beauty, which is extraordinary. Yeah, that's kind of the vision. That's where I'd take you guys to.'

This imagined city becomes slightly more vivid as he speaks. 'I think within that reach, I can sort of describe the complex, rich flows of transactions that are going on because these ecologies have become sort of self-feeding in beautiful ways. And this could be even just a small street seeding all these different ecologies. I think it's a symbiosis, like some sort of cafe or restaurant, that offers a density of micro relationships. They however take time to grow…and it requires the learning capacity of those other agents to be able to move and to dance with each other. I can't quite describe it, but I can start to imagine it.'

It's not mere futurism but a working vision of what might be possible if we could fully unleash the distributed agency that Indy sees as already present in our world, waiting for the right conditions and frameworks to flourish in ways that could transform how we understand governance, economics, and the fundamental organisation of human settlements.

This vision emerges from his understanding that our current urban challenges—from climate change and inequality to the breakdown of democratic institutions—are symptoms of deeper systemic failures that require new approaches to how we organise collective life. Rather than simply building more efficient cities or implementing better policies within existing frameworks, Indy argues for what he calls 'institutional innovation' that can create new forms of collective intelligence and distributed decision-making.

The seeds of this vision were planted during his architectural studies at the University of Bath, where a pivotal encounter with a book about labyrinths and pyramids would fundamentally reshape his understanding of power and possibility. Sitting in the university library, the young Indy encountered an idea that would become central to his worldview: that the pyramid represents transparency and asymmetric power, whilst the labyrinth embodies intimacy and freedom but also the power of obscurity. 'It pivoted my worldview,' he recalls, 'from actually having grown up in a world of conversation of modernity and radical transparency ... [to] realising [that] the asymmetric power story was a power base summit. Power is fundamentally, you know, a world constructed with vast asymmetries.' (Fig. 11.2).

This insight—that transparency could actually aid in the acceleration of asymmetric power—became a foundational understanding that would inform decades of work at the intersection of systems change and social innovation. The metaphor of labyrinth versus pyramid has become central to Indy's thinking about how to design systems that enable genuine participation rather than mere consultation and that create space for emergence rather than imposing predetermined outcomes. The pyramid, with its clear hierarchies and transparent command structures, appears efficient and accountable but actually concentrates power whilst creating the illusion of democratic oversight.

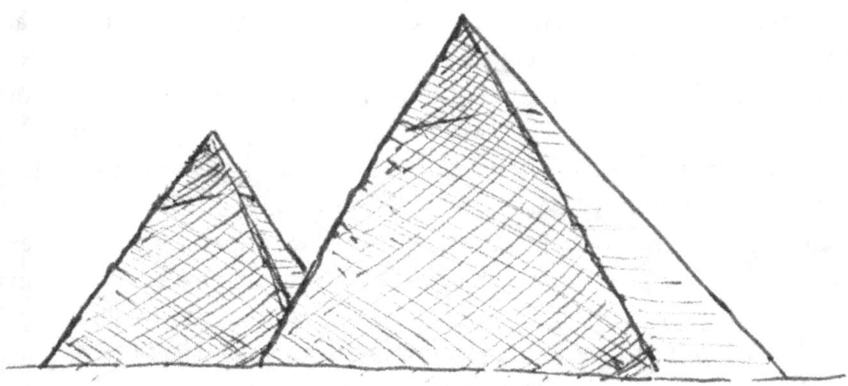

Fig. 11.2 An illustration of pyramids, mentioned by Indy Johar as a seed of understanding of power and possibility

But theory alone did not shape Indy's perspective; his understanding of power and collective action was profoundly influenced by witnessing the 1984 riots in Delhi, where he saw firsthand how majorities could be constructed and instrumentalised by political actors seeking to advance particular agendas. This experience left him with a deep scepticism of simple narratives about collective action and state power. The riots revealed how quickly social fabric could be torn apart when political entrepreneurs found ways to weaponise difference and how communities that had coexisted peacefully for generations could be turned against each other through strategic provocation.

'What does it mean to build societies of freedoms of care rather than freedoms of consumption, where the freedom to be radically human is rooted in those models? If we can unleash the freedoms of care from all of society, is that not profoundly more significant than the coalitions of power that we tried to construct? To see a new position beyond the kind of libertarian hyper-rational individualism. One of radical freedoms to care … that recognises us as relational through our relationality rather than our collectives. And what would it then mean to build a city which is rooted in those deep relationalities rather than in our theories of collectives?'

These wonderings have inspired much of his work since then, leading to the creation of Innovation Hub networks, social investment structures, and eventually Dark Matter Labs, where he continues to explore how multiple disciplines and possibilities can be brought together through an approach that is deliberately intersectional. 'It's not the diversity,' he emphasises, 'it's the intersection of the disciplines.' For Indy, real innovation happens when we can think simultaneously about 'architecture without thinking about material constraints…material constraints without understanding carbon constraints…carbon constraints without thinking about labour constraints.' It is in these intersections that new possibilities emerge, requiring thinking across multiple scales and systems simultaneously rather than treating challenges as isolated problems.

This way of thinking has made him increasingly critical of traditional academic approaches to knowledge production, where, 'unfortunately, academia are so siloed that it doesn't deal with intersectional knowledge,' leading to the observation that 'most of that innovation coming out of academia is actually without utility, because it advances propositions that don't understand its intersectional realities with other disciplines.' The university system, with its departmental boundaries and disciplinary gatekeeping, struggles to address challenges that by their nature require integration across different domains. His intellectual nomadism, or 'intellectual promiscuity,' as he himself describes it, is not mere dilettantism but a deliberate strategy for understanding complex systems in their fullness. 'So I don't blame academics or academia. I think it's just the nature of the shift of our knowledge system. So it is really vital that we continuously document the intersection of knowledge and share this in order to build the field.'

His process of knowledge creation has evolved with technology, where he speaks of dictating thoughts directly into ChatGPT, using it as a tool for rapid iteration and refinement of ideas. 'I use it like a thinking partner,' he explains, describing how the technology enables him to test ideas quickly, to explore different framings of problems, and to generate multiple approaches to complex challenges.

What makes Indy's approach unique is his emphasis on what he calls 'pattern anomalies' rather than pattern affirmation. 'The most interesting people I meet are constantly looking for pattern anomalies,' he explains. 'The most boring people I meet are looking for pattern affirmations.' This orientation towards anomaly rather than confirmation allows him to maintain intellectual freshness, to identify weak signals that might indicate emerging possibilities, and to challenge assumptions that have become so naturalised they are no longer recognised as choices.

This approach extends to his relationship with ideas themselves, where 'the things you own own you,' emphasising the importance of maintaining a certain detachment from even our most cherished concepts. This philosophical stance allows him to remain open to new possibilities whilst still engaging deeply with transformative ideas, to hold strong convictions whilst maintaining intellectual humility.

> 'To be able to be at the intersection of philosophy and doing is—I think—a superpower in this modern world. And to be able to move from the super abstracted to philosophy and then to craft, and be able to move between

those scales, I believe is really, really vital. I remember I wasn't a great student for my first two years. Until my teacher turned around and said, write me an essay on the future of the city. Then suddenly the world clicked. The capacity to be able to self-author to operate between technique and philosophy. That I think is actually the critical capability of the twenty-first century.'

When asked about the team he would assemble to rebuild our society after collapse, Indy's first response is slightly unexpected: patience. 'Not saying too much too soon. Let a system say what it wants to say,' he advises. This approach—of waiting for deeper energies to emerge after initial reactions have played out—reflects his understanding of how complex systems actually transform, requiring what he calls extraordinary leadership in every location rather than relying on individual heroes.

'I think the individual hero is a rather boring place. The sort of team I'd like to be part of demonstrates extraordinary leadership, because I think I enjoy seeing brilliance. Sometimes I am disappointed in myself and others if we just deliver a project the way we have imagined it. A waste of time. The point is to be better, be different, be situationally more appropriate than any preconception you could bring to the table. I love that. The contribution of the situation … considered in different ways.'

Recircling the conversation to how this all plays out in our cities, Indy builds on his train of thought, suggesting that the true power of a city lies not in its moments of heroism but in 'how it holds fragilities.' This insight cuts against conventional urbanist wisdom, suggesting that a city's greatness might be measured not by its monuments or achievements but by its capacity to create space for vulnerability and care. He frames how cities fundamentally are about relationships rather than infrastructure, about the quality of encounters between different people and communities.

As our conversation draws to a close, we ask whether he can spot pockets of the future. He allows a moment of silence, after which he shares Marshall McLuhan's insight that 'when you flood people with information, they stop processing.'[1] Instead, people begin pattern matching, leading to a situation where politicians must either violently break existing patterns or conform to them, creating polarisation and simplification. Yet even in this challenging context, Indy remains hopeful about the possibility of new forms of communication and leadership emerging, speaking of 'spell words'—terms that carry deep transformative power—and suggesting that we are approaching a moment when new forms of oratory might emerge to meet our present challenges.

[1] McLuhan, M. (2017). The medium is the message. In *Communication theory* (pp. 390-402). Routledge.

12

Jess Miller (Australia): The City as a Canvas

Abstract Councillor Jess Miller is Deputy Lord Mayor of Sydney, focused on delivering innovative outcomes combining sustainability, economic inclusion, creativity, and affordability. Her environmental awakening began among the Mornington Peninsula (Victoria, Australia) mangroves, while an early São Paulo visit displayed urban systems 'operating at the absolute boundary of system limits.'

Keywords Urban climate action • Local government • Stakeholder engagement • Sustainable urbanism

12 December 2023, Online from Sydney, Australia; Updated December 2025

The first thing one would remark about our conversation with Jess Miller is that it has the vibe of a high school reunion or campaign meeting (Fig. 12.1). She jumps into the talk as if we are picking up on a long-running conversation about the new urban order that we had left only a

C. Oke, J. Verbeeck, *City Leaders for Sustainable Futures*,
https://doi.org/10.1007/978-981-95-9544-0_12

Fig. 12.1 Jess Miller, drawing inspired by photograph by Nic Walker

handful of days ago. Yet she is not the type of person to just fill time. She does not want to waste it. There are things to be done. That is also what her body language conveys. She reads our minds as we observe her. As our conversation draws to a close, Jess reflects on what sustains her through the long work of urban transformation. 'You know, once you get it, you can't not get it,' she reflects. 'And once you've seen it, you can't unsee it.' This is not a burden but a calling: the recognition that understanding systemic crises creates both responsibility and opportunity and that seeing clearly what is wrong also reveals possibilities for making things right.

Radiating the easy confidence of someone who has spent years moving between boardrooms and protest sites, between community gardens and council chambers, Jess settles into our conversation from her Sydney home. Re-elected to Sydney City Council at the end of 2024, and currently the Deputy Lord Mayor, she demonstrates the kind of urban leadership that emerged from the grassroots, was forged in the crucible of direct action, and now operates within the formal structures of local government whilst never losing its irreverent edge. She is the type of urban practitioner for whom urban policy becomes a form of creative practice, community organising becomes performance art, and the machinery of local government a canvas for reimagining how we might live together differently. Her journey from the mangroves of Western Port Bay to the chambers of City of Sydney Council illuminates how individual transformation and urban change intertwine and how personal awakening to environmental crises can turn into collective action for systemic change.

Her environmental journey began in during her childhood in the mangroves and seagrass beds of Victoria's Western Port Bay, where her mother took her to witness both natural beauty and its destruction. 'I do remember Mum taking us to this quite magical protest, where these artists had built these incredibly huge puppets looking like ibis and seabirds, trying to save bushland from developers. I remember being very taken by that and then incredibly confused and sad and angry—you know, powerless little kid anger about why anybody would destroy the land and life that was there (Fig. 12.2).'

The scene, combining art, environmental protest, and community gathering, would become a template for her vision of how change could

Fig. 12.2 Seabird puppet illustration, mentioned by Jess Miller at an early moment of her environmental journey

be made tangible. As a combination of beauty and urgency, of creativity and politics, and of individual emotion and collective action. The formative influence was her mother, a young nurse who had become an unlikely environmental activist, dragging her children to protests and bushland restoration projects not because of any grand political ideology but from a simple conviction that it was 'the right thing to do.' 'So there were weekends where my mum made me and my brother go find swamp rat scats because they were an endangered marsupial and their habitat was under threat of development. We spent some Saturdays looking for rat poo. And even better than that, she then sent us off to school on Monday with specimens in our lunchbox wrapped up in [plastic] wrap for show and tell.'

This early experience of activism was characterised by what Jess calls 'working-class activism'—which was a very different type of activism born out of a sense of ownership and desire to protect beautiful places,' informed not by university theory or environmental science but by lived experience of specific places under threat, a sense of duty, and the visceral understanding that some things are worth fighting for simply because they exist and are beautiful and sustaining.

The transition from childhood witness to adult activist came through her involvement with the clean ocean campaigns at Gunnamatta beach on the Mornington Peninsula near Melbourne, where local surfers were confronting the reality of raw sewage being pumped into the water where they surfed. In these campaigns serious health concerns were combined with the irreverent humour that would become characteristic of her approach to activism. She smiles infectiously while describing the image of surfers in wetsuits dragging barrels of effluent up the steps of state parliament during a sitting week. It perfectly captures her idea of effective protest: visceral, visual, impossible to ignore. 'One by one, they just let it rip. So all of that water covered the steps of parliament during a sitting week. So the point was made incredibly well, the smell of it, the look of it,' she recalls. This was not symbolic action but a tangible demonstration of the consequences of policy decisions, forcing politicians to confront the physical reality of what their decisions meant for real places and real people. 'You can't just care and believe in it. You've got to do it with your mates, the people you like, and find some humour in it.'

Perhaps like the Brazilian city itself, Jess gravitates towards the edges, the extremes, and the places where systems are most visibly strained and therefore most clearly visible. In São Paulo, she encountered what she describes as a place that was 'operating at the absolute boundary of system limits'—where the contradictions of urban development became visceral, where inequality became architectural, and where the environmental crisis became an everyday reality that millions of people had to navigate. If not online, we would meet her not in Sydney's gleaming harbour but in São Paulo, a city that has haunted Jess's imagination for more than two decades.

It was there that as a young woman from Somerville, just four days shy of her 18th birthday, she first encountered what she calls 'a real city'—a place of such overwhelming scale, complexity, and contrast that it fundamentally shifted her understanding of both urban possibility and urban peril. 'It was the first city, I guess, the first real, real city, I think I'd ever been to outside of Australia, really. It was like nothing I could ever have even imagined in terms of the size of the city. The smells of the city. The contrast between the haves and the have nots. Though the thing that struck me most is how the city is characterised by the Tietê River. And the smell and the condition it came with. This suffocating stench. I couldn't really grasp how it had gotten to that point, that a system can be so completely out of balance. And yet, it's still one of the most brilliant places I've ever been to. Being so big and vibrant.' A Rotary exchange programme had transported her from the familiar rhythms of suburban Melbourne to the pulsing vastness of Brazil's largest metropolis. It was an existential confrontation in which she recognised that urban systems could become so completely out of balance that the very elements that should sustain life become toxic. This early encounter with urban crisis at scale would profoundly shape her understanding of activism and change. São Paulo exposed possibilities that would stay with her for decades. When she returned with her own family years later, she found that Avenue Paulista—that massive arterial street she remembered as choked with traffic and fumes—had been transformed by a separated cycleway running down its centre. 'I just thought, "Wow, if you can do it there, you can do it anywhere. There's just no excuse to not be able to cycle if you can do it down that street."' An otherwise simple infrastructure investment had become a proof of concept. That image of reimagining and transforming even the most challenging urban environments sets the scene for the rest of our conversation.

After her transformative experience in Brazil, Jess had returned to Australia with a fundamentally altered perspective on both art and activism. She had been accepted to study painting at art school, but her encounter with São Paulo's urban crisis left her feeling she had 'nothing worthwhile to say' and did not 'understand how the world works.' How could art and activism go together? Her move to Sydney brought her into contact with Holly Creenaune, a 'grassroots activist' living in a squat rigged with solar panels, practising a kind of integrated politics of direct action and alternative living. 'Whatever Holly was doing, I got involved with, and it was fun,' Jess remembers. This period saw the reinvigoration of Friends of the Earth Sydney, characterised by a combination of serious climate activism and what she calls 'punk' energy that did not fit neatly into established environmental organisations. She suddenly raises her voice. 'Permablitzes! You know, turning a really crappy garden or backyard into a beautiful garden—an afternoon with music and food and beers. It's permaculture but then combined with a backyard blitz. It was based on reciprocity,' Jess explains, describing her role in organising sites, connecting people with skills, and creating the conditions for collective transformation. It was not enough to oppose destructive development. Activists also needed to demonstrate what transformation felt like.

Throughout this period, Jess played a game with her activist friends called 'fantasy front page': imagining what environmental activism might look like if it were front-page news and part of mainstream culture. As part of communications teams for various campaigns, she would 'relentlessly pitch articles to *Vogue*' and dream of a front page featuring both environmental action and cultural figures like footballer Brad Miller from the Melbourne Demons. It was more than naive optimism; it was an understanding that cultural change is as important as policy change and that shifts in consciousness often precede shifts in behaviour. Years later, she would see vindication of this approach when Emma Watson appeared on the cover of *Vogue* for a sustainability issue, when rugby player David Pocock defended his right to protest and went on to become a senator. 'If you're in it long enough, it's a real source of hope,' she reflects. 'To see how being on the progressive edge, after a while you push and push and push, and then you can step back and there's momentum built behind you.' Using culture as a site of political transformation would inform her later work with the agency Republic of Everyone, where she develops strategic communications and behaviour change programmes through 'Trojan horsing': putting the behaviour change within something people wanted, not something they were guilted into doing. She mentions projects like Garage Sale Trail and Grow It Local, which offered people ways to participate in more sustainable practices without requiring them to first embrace environmental ideology.

'Toppling capitalism is really scary because no one can tell you what's going to replace it,' she observes. 'So if you can package up alternatives in a way that feels generous and fair and desirable, then great.' In recent times she has been working on a New Normal Sydney, with the 'new' representing what she calls 'a culmination of all that learning.' 'It is like an IKEA of all the solutions that we need.' It is about making existing solutions more accessible, more attractive, and more implementable. The barriers to sustainability are often practical and cultural, rather than technical. We all agree that the metaphor of IKEA is particularly apt, in that it suggests solutions that are modular and accessible and come with clear instructions for assembly, acknowledging that even the best solutions require some work to implement, and that transformation is always a collaborative process between those who design systems and those who use them.

Jess's approach to activism has been profoundly shaped by her longtime practice of Ukidokan, a style of martial arts, which she credits with providing both humility and resilience. 'Just when you think you get it, you don't get it,' she says, describing how the practice constantly reminds her that mastery is illusory and that there's always someone 'taller than you and faster than you and younger than you' who can challenge your assumptions. 'How can we adapt under pressure?' she asks. 'That is going to be the single greatest quality of human beings for the next foreseeable decades.' When asked who needs to be in the Trojan horse, she outlines what she calls 'genres of people' rather than specific individuals: 'A really good project manager, somebody who's really rich and has a lot of land, a technical expert, and somebody with a connection to culture. Ideally, that rich person would be tragically in love with an artist or creative and willing to be inspired by that person,' she adds with trademark wit, acknowledging that even necessary compromises with existing power structures can be humanised and redirected. 'Ideas are cheap,' she will tell us at the end of our talk. 'If you let these people loose to experiment and do stuff, I think that's the most important thing. There's not enough respect or acknowledgement or humility that can be handed to the "doers" to enable them to get on with it.'

Her source of hope lies not in technological solutions or policy reforms but 'the brilliant people doing stuff'—the networks of practitioners, innovators, and organisers who are creating alternatives in cities around the world. 'Sometimes it's a little bit heartbreaking that nobody else knows about it,' she says, speaking of the gap between innovation and recognition, between local transformation and broader awareness. She sees cities as simultaneously spaces of enormous possibility and significant constraint and places where new forms of living can be prototyped but also where existing power structures resist transformation, where social and environmental visions can be made visible, and where the abstract work of policy and planning becomes a concrete experience for millions of people. The canvas metaphor also suggests that urban transformation is always creative work that requires imagination, aesthetic sensibility, and the ability to see possibilities that do not yet exist, explored by people working both inside and outside formal structures to challenge existing arrangements and create attractive alternatives.

13

Carlos Moreno (Colombia/France): The Mathematics of Human Connection

Abstract Carlos Moreno is a French-Colombian multidisciplinary scholar, Professor at IAE Paris–Sorbonne, internationally recognised for pioneering ideas that enhance urban life. Carlos proposed the 15-Minute City in 2015, asking: 'Why do we need to spend a lot of time commuting? Why do we have this single use of buildings?'

Keywords 15-minute city;·Compact city • Sustainable mobility • Social sustainability

31 January 2024, Online from Paris, France

With the winter sun streaming through his Parisian office window, Carlos Moreno shows a living laboratory of urban transformation as he adjusts his camera to capture the protected bike lanes that now flow like rivers of sustainable mobility where once only cars dominated. Beyond his window, the sounds of contemporary Paris create a symphony that would have been unimaginable when he had first arrived as a political refugee from Colombia: the gentle whir of electric bikes, the conversations of

C. Oke, J. Verbeeck, *City Leaders for Sustainable Futures*,
https://doi.org/10.1007/978-981-95-9544-0_13

pedestrians reclaiming street space, the absence of the traffic jams that had once stretched for 200 kilometres daily across the city—all testament to the transformation he has helped design through what the world now knows as the 15-Minute City concept, though for Carlos it represents something far deeper than urban planning theory (Fig. 13.1).

If Carlos could choose our meeting place, the answer comes without hesitation: we are already there, in his beloved Paris, at this precise moment in winter 2024 when the city stands as proof that radical urban transformation is not only possible but inevitable when guided by principles of human dignity and environmental necessity. 'My favourite place for this interview is in my city, in Paris. The best moment for this interview is today,' he declares with the conviction of someone who has watched impossible dreams become concrete reality beneath his office

Fig. 13.1 Carlos Moreno, drawing inspired by photograph by Thomas Baltes, supplied

window. That view tells us without words that transformation does not happen in abstract planning documents but in specific places, at particular moments, through the accumulated actions of people who refuse to accept that the way things are represents the way things must remain. Paris welcomed him as a refugee and became the canvas where his mathematical training, South American roots, and European opportunities could be combined to create new possibilities, to help rethink how millions of people might live together more sustainably, and to demonstrate how individual journeys of displacement and discovery can contribute to collective transformation that extends far beyond any single life or career.

From his window he can trace the visible evidence of transformation that began with ideas but became reality through political courage and citizen engagement. 'When I arrived as a political refugee, I was just 20 years old. I embraced the city the way it embraced me, and many years later I had an incredible chance to be involved in its radical transformation,' he says, his voice expressing both gratitude and determination. The view captures everything he believes cities can become: the traditional architecture of the French Academy across the Seine blending seamlessly with contemporary interventions, the river serving as both historical constant and modern recreational space. Protected bike lanes create safety for sustainable mobility, and urban parks replace parking spaces to provide green refuge in the dense urban fabric. 'This city really has this capacity for blending the traditional architecture with modernity, to ensure Paris is not a museum, but a city that is liveable,' he explains, capturing his philosophy that preservation and innovation must work together rather than against each other and that cities must honour their past whilst adapting to contemporary needs and future challenges.

He always says this with words imbued with deep gratitude towards the person who listened to him and trusted him: the Mayor of Paris, Anne Hidalgo. A pioneer, committed, determined, and courageous, she is someone for whom he has boundless admiration. He describes her as the best mayor in the world. He highlights her ability to listen, to trust ideas before they become fashionable, and to transform vision into action. Through political courage, consistency, and perseverance, she has made profound urban transformation possible. For him, her confidence has been decisive, giving legitimacy, continuity, and strength to a people-centred vision of the city, grounded in proximity, social justice, ecological transition, and everyday quality of life, that can inspire cities beyond Paris. His respect goes hand in hand with loyalty and a shared conviction that cities can change lives.

A series of photographs Carlos chooses to represent how he thinks about his sustainability journey carries profound personal and political resonance, connecting his current work in European cities to formative experiences with Indigenous communities in rural Colombia. In one photograph children and adults wearing traditional clothing, their relationship with land and water and natural cycles visible in every detail of their daily lives, representing what he calls 'the first time in my life to discover the importance of sustainability.'

'In the 70s, when I was a teenager in Cali, it was the third largest city of Colombia and a very intellectual city, bustling with a lot of students, intellectual writers, etc. I grew up mostly in the city, as our family, who were small farmers, had been impacted by processes of violence and land grabbing. It left them no other choice than to move to the city, even if they had been farmers for as long as they could remember. So it was only around the age of 13 that I discovered there were a number of traditional Indigenous cultures quite close to my city. West of the city, there was, for example, an important Indian minority, which today is referred to as the Misak people. It refers to the Indian word of this province, but as such it's not really a word that existed [in Spanish]. It was an Indian word. I was totally impressed by their lifestyle. In the photo, you can see the traditional clothing of the women and men of the tribe. It is how they still live today. In their daily life, there is a close relationship between the community and nature. Kids are taught to respect nature, water, air, sun, and rain. But also respect for collective work and thinking, organised via so-called Mingas.'

The exposure to Indigenous knowledge systems opened his mind to possibilities that formal education had never presented, demonstrating how childhood encounters with different ways of living can plant seeds that bloom decades later into approaches that challenge dominant paradigms about development, progress, and the relationship between human communities and natural systems. Carlos describes the painful transition that forced rural people into urban spaces without adequate support or understanding of how cities might serve their needs. Yet witnessing how the Misak community maintained its relationship with natural cycles provided him with a template for thinking about sustainability that transcended the rural-urban divide, suggesting ways to integrate traditional wisdom with contemporary urban challenges. 'The visits to this community I think have opened for the first time my mindset towards this natural sustainability.' (Fig. 13.2)

The journey from these revelations as a 13-year-old boy to his current role as one of the world's most influential urban thinkers required crossing multiple intellectual and cultural borders. He was 20 when he arrived in Paris as a political refugee. Studies in Paris would lead him into mathematics and computer science and subsequently to the application of technology to urban challenges. But because he lacked a professional training in urban planning or architecture, his approach emerged mostly from lived experience. His early career focused on using mathematical and computer science tools to optimise infrastructure and reduce traffic congestion, positioning him as a pioneer in what would become a massive global industry. Yet something fundamental was missing from the technological optimism that dominated smart city thinking.

The breakthrough came through encountering the work of French sociologist and philosopher Edgar Morin, whom Carlos describes as 'the universal father of complexity.' Morin's complexity theory provided the

Fig. 13.2 An illustration of one of Carlos Moreno's photos that represents his sustainability journey

intellectual framework for understanding cities as living organisms rather than mechanical systems, emphasising interdependencies rather than isolated optimisation, horizontal connections rather than vertical hierarchies.[1] 'Embracing complexity means to work together, to develop an in-depth approach on the interdependencies, to design the model to give importance to the link between the different systems instead of considering one system separated [from] the other systems,' Carlos explains, describing how this paradigm shift fundamentally changed his understanding of what cities could become.

The influence of complexity thinking intersected powerfully with his recognition that the technological solutions he was developing primarily served wealthy European cities whilst failing to address the fundamental challenges urban populations were facing in the Global South, where 'the question of the technology is totally different because the real question is the access to essential services like access to water, access to education, access to culture, access to health, etc. The question of the traffic jams is not the main question.'

This intellectual renewal led to a crucial encounter with Muhammad Yunus, the Nobel Prize-winning economist whose concept of 'three zeros'—zero carbon, zero poverty, and zero exclusion—provided a framework for thinking about urban development that could serve everyone, rather than just the privileged.[2] 'This vision was a new equation for me, a common point for transforming cities in both the Global North and the Global South,' Carlos reflects, describing how Yunus's work, in line with that of Edgar Morin, suggested possibilities for urban transformation that addressed environmental and social challenges simultaneously rather than treating them as separate issues requiring different solutions.

The convergence of complexity theory, global equity concerns, and practical experience with urban technology created the intellectual foundation for what would become 'the 15-Minute City' concept, yet the path from theory to implementation required another crucial encounter: with the emerging global network of cities taking leadership on climate

---

[1] Morin, E. (1992). From the concept of system to the paradigm of complexity. *Journal of Social and Evolutionary Systems*, *15*(4), 371–385. https://doi.org/10.1016/1061-7361(92)90024-8

[2] Yunus, M. (2017). *A world of three zeros: the new economics of zero poverty, zero unemployment, and zero net carbon emissions*. Hachette UK.

action. Through Carlos's involvement with C40 Cities, initially created by London Mayor Ken Livingstone and later expanded through partnerships with Michael Bloomberg and the leadership of Anne Hidalgo, he gained access to mayors and urban leaders worldwide who were seeking practical approaches to reducing emissions whilst improving quality of life for residents.

The breakthrough moment came in 2015 during COP21 in Paris, when the intersection of global climate negotiations and Paris's urban vision created space for new approaches. 'I proposed for the first time "the 15-Minute City" as a new trajectory for naturally reducing our carbon footprint via electric cars in the centre of the cities.' He describes how the concept emerged from the practical recognition that technical solutions alone would never achieve the emission reductions necessary to address climate change.

The 15-Minute City concept addresses what Carlos identifies as fundamental inefficiencies in how cities organise space and time: 'Why do we need to spend a lot of time commuting? Why do we have this single use of buildings?' These and other questions, posed during a moment when climate urgency was creating political space for radical experimentation, suggested a different approach to urban development: one that prioritised access over mobility, mixed use over specialisation, and local services over distant consumption.

Yet Carlos emphasises that the 15-Minute City is not a universal template: 'I propose the 15-Minute City as the calculus territory that is valuable for cities both in the Global North and the Global South, even though the implementation needs to be tailored to different needs.' This recognition reflects his understanding, rooted in complexity theory and cross-cultural experience, that sustainable urban solutions must emerge from local conditions rather than imported models.

A number of urban pioneers, from historical figures to contemporary practitioners, most profoundly shaped his understanding of how change could happen in cities. Each of these pioneers have offered distinct insights into the relationship between ideas and implementation. Jane Jacobs holds special significance as 'the mother of all urban battles,' someone who demonstrated that effective urbanism requires both intellectual rigour and practical engagement and both critique of existing systems and creation of alternatives that serve human needs rather than abstract planning principles.[3] 'It's not enough to be a thinker. We need to be doers,' Carlos says, reflecting on Jacobs's work and influence.

Contemporary figures who embody this tradition include Jan Gehl, whose work on cities for people provides practical methods for humanising urban spaces; Jeff Speck, whose advocacy for walkable cities demonstrates how technical interventions can serve broader social goals; and a new generation of practitioners, like Brent Toderian and Carmen Santana, who are adapting these approaches to different cultural and geographical contexts.

His work and thinking also carry a spiritual dimension. His Colombian roots are apparent, but encounters with leaders like the Dalai Lama and Lobsang Tenzin whose influence extends far beyond their specific political contexts are also incorporated into his approach. 'If we need to humanise our cities, we first need to find our inner humanity again. This means to be at the same time aware that we need to be humble and to develop more otherness,' Carlos says. He elaborates on how sustained engagement with questions of justice and sustainability requires not just intellectual understanding but emotional and spiritual development that enables a deeper connection to both human and natural communities.

This emphasis on inner transformation as a prerequisite for outer change reflects his understanding that cities become sustainable not through technological optimisation but through cultural transformation that enables people to see themselves as part of larger ecological and social systems, rather than separate from them.

The practical application of these insights becomes visible from Carlos's office window, where the protected bike lanes represent not just infrastructure investment but a cultural shift towards prioritising human needs over automotive convenience, whilst the transformation of parking spaces into urban parks reflects not just environmental improvement but political courage to challenge entrenched interests in the service of collective well-being. The integration of traditional and contemporary architecture demonstrates not just aesthetic preference but recognition that sustainable futures must build on, rather than replace, cultural heritage,

[3] Jacobs, J. (1961). *The death and life of great American cities* (pp. 50–54). Random House.

creating continuity between past and future that honours both tradition and innovation.

These visible changes in Paris represent proof of concept for approaches that can be adapted to cities worldwide, yet Carlos remains realistic about the challenges involved in scaling transformation. 'In face of big challenges like climate change, we today need to go to the essentials of a society: rethinking the economic productivity, the irrational consumption, the social inequities and ultimately social exclusion.'

His optimism about possibilities for transformation emerges from a mathematical understanding of how complex systems change—not gradually or predictably but through sudden shifts when multiple factors align to create new stability.

14

Mandy Nicholson (Australia): Listening to Country

Abstract Dr Mandy Nicholson is a Wurundjeri artist, Woiwurrung language specialist, and Traditional Custodian of Naarm (Melbourne). She also has connections to Djaara and Ngurai Illum Wurrung people of Victorian and Irish and German heritage. She founded the Djirri Djirri dance group in 2013, and for over 30 years, she has created culturally grounded works through Tharangalk Art.

Keywords Indigenous knowledge • Decolonial planning • Urban governance

28 February 2024, Online from Melbourne, Australia; Updated December 2025

'We've definitely lost the ability but not completely lost the gift of seeing and hearing. We've just forgotten how to tune out from all the city silliness and to focus on things. Hearing birds and nature sounds and taking our shoes off is medicinal. We forgot how to do that. I always tell everyone how even right in the middle of the city you can tune your ears into

C. Oke, J. Verbeeck, *City Leaders for Sustainable Futures*,
https://doi.org/10.1007/978-981-95-9544-0_14

Fig. 14.1 Mandy Nicholson

the bird sounds and you'll hear a tiny little parrot above all the city noise. You might hear or see the black cocky [cockatoo] in the distance, and it will tell you that it's going to rain. You just have to adapt to how you're living to remember Country.' (Fig. 14.1)

After delivering a three-hour lecture titled 'Engaging with First Nations People in Design,' Mandy Nicholson appears on screen for our conversation with the gentle satisfaction of work well done, a map of Australia displaying the 268 distinct First Nations language groups across the continent in the background. It is a visual reminder that what settlers call 'Australia' encompasses territories as diverse and culturally complex as all the countries of Europe combined. The map behind her does not just show linguistic diversity; it represents thousands of years of sophisticated land management, social organisation, and spiritual practice that European colonisation sought to erase but which persists in the knowledge, stories, and continuing presence of Aboriginal and Torres Strait Islander people.

It is therefore not a surprise that Mandy opts for a small creek in Healesville called Coranderrk in Woiwurrung language—known in English as Badger Creek—if we had had the opportunity to meet in person. It is near the site of an Aboriginal station established in 1863, where Wurundjeri people and other Aboriginal groups from across Victoria were forced to live under government control. Yet despite the traumatic history of institutional control and cultural suppression, the land itself holds deeper stories—stories of sustainable living, sophisticated land management, and spiritual connection that predate colonisation by tens of thousands of years. The creek provides what she calls 'cultural recharge,' where sitting beside flowing water and listening to native birds remind her of the larger purpose underlying her work: maintaining connection to Country even within urban environments that often seem designed to sever such connections.

'I live about an hour away from there. But I always go back as much as I can. To go pay my respects to my loved ones and to connect to a different part of life that not everyone gets to enjoy because they're so wrapped up in their crazy busy agenda. They forget to stop, take a breath, close their eyes, and feel the grass between their toes. And how it feels to have the wind in their hair and just pause for a minute. From the creeks in the valley area, I'd go up to the mountain area, and I would sit underneath the mother trees and just be still. Ninety-nine percent of our old-growth forest has been logged, but there's one small patch with a number of old-growth forest trees, which helps me feel the vibrations of the earth. The vibrations of everything around me: the bugs, the birds, the little gusts of wind, the mottled sun that comes through the branches of the trees, and the silence as well. I wonder about how much these trees have seen happen in their lives.'

She shares her thoughts on how the tree has witnessed the transformation of landscapes and communities. Yet rather than dwelling on loss, Mandy uses the tree's survival as a teaching tool for understanding how cultural knowledge persists even through catastrophic disruption.

'Imagine the tree being chopped down—that represents invasion. But the roots represent language, and they grew deep and kept that trunk alive. The trunk represents the strength of our elders and their knowledge, giving us that cultural foundation.' The metaphor extends to describe how cultural renewal happens across generations, where 'the shoots that come out because the roots have kept them alive—that's the community leaders that are out and about doing as much as they can to help culture be present and visible and help those elders.' These emerging leaders, nurtured by ancestral knowledge but adapted to contemporary conditions, create bridges between traditional and modern worlds in an ongoing cycle where 'over the hundreds of years of life that she's lived, this mother tree has dropped lots of seeds, and the seeds represent the babies and the next generations coming. Eventually, when the mother tree is at the end of her natural life, she's still noticing these little ones coming up through the forest floor and notices [that] these little ones come back up, and they become the strong mother trees themselves, and that cycle can continue.'

This vision of cultural continuity as a natural process—requiring both protection of existing knowledge and cultivation of emerging leadership—shapes all of Mandy's work, from language revival to dance performance to academic research, providing a framework for understanding how ancient cultures adapt to contemporary challenges whilst maintaining essential qualities that connect people to place, to each other, and to responsibilities that extend across generations.

A recent pre-dawn visit to the mother trees provided what she describes as 'probably one of the most special moments in my whole life.' She had arrived before sunrise in preparation for filming work to find herself alone in the forest as dew created 'little clouds above the tree canopy.' In the profound silence of that early morning, she began to hear what seemed like a dawn chorus of birds: 'currawongs, corellas, cockatoos, really all the different types of parrots,' until she realised, 'There were no birds in the trees. So where's this noise coming from? And it was all the lyrebirds.' These remarkable creatures, found only in Australia, can perfectly mimic any sound they hear—from other birds to camera clicks to chainsaws. Like the lyrebird, Mandy's work involves careful listening, precise reproduction of cultural elements, and creative adaptation that maintains authenticity whilst enabling connection across different worlds (Fig. 14.2).

Fig. 14.2 A lyrebird, as mentioned by Mandy Nicholson

This encounter reinforced her understanding that cultural work requires developing what she calls 'cultural ears and eyes': the ability to hear and respond to the multiple layers of communication happening in any environment. In urban environments, this means learning to hear Indigenous presence in places where it has been rendered invisible, learning to recognise traditional knowledge in contemporary practices, and learning to understand how ancient relationships with Country persist even within built environments that seem designed to sever such connections.

The journey to this understanding began in childhood 'on Country' but was interrupted by 'awkward teenage years' spent in Gippsland, 'off Country.' This for her was 'a culturally unsafe environment' where displacement created gaps she recognised but could not yet address. High school proved to be a racist environment where 'I hid away from my identity because I wasn't allowed to be who I was,' leading to a transformation that began at 18 with enrolment in a Koori Studies course at Monash University Churchill campus, a programme designed specifically for Aboriginal students making the transition to higher education.

What triggered real transformation was being asked to write about her life. 'I wrote down everything that I knew about myself—I knew all my bloodlines, my land, my country, my connections on both sides of my black family.' The act of documenting her knowledge revealed its richness and completeness, challenging the shame that racist environments had imposed. 'Why am I hiding from all of this?' she wondered. 'I'm very lucky to be able to know this.' She continues, 'That was the day I decided to do as much as I can to help bring back culture—for my future kids if I was ever to have kids, which I ended up having two, and for my community, because I've seen [during] my whole life the young ones just get lost, and they've got no cultural foundation to fall back on.'

Working with language was a first step in that journey, 'translating nursery rhymes, then children's books,' understanding that language revival must start with the youngest generations. Yet the work quickly evolved into sophisticated analysis where she could 'break words apart and find out the different meanings within one phrase and get extra cultural information that's been hidden for such a long time.' It provided insight into how Indigenous languages encode sophisticated understanding of ecology, social relationships, and sustainable living practices that remain relevant for contemporary challenges.

The creation of the Djirri Djirri dance group represents perhaps her most visible contribution to making Indigenous culture present in urban environments through public performances that combine traditional knowledge with contemporary presentation. The group addresses what she identifies as a fundamental problem in urban Australia: 'If you are somebody that's just visiting Melbourne for a day, or someone that's lived here their whole life, they probably would come up with the same answer—when asked where the Aboriginal people live, where the First Nations people are. They can't see us.' This invisibility is not accidental but designed, the result of policies and practices that systematically removed Indigenous people from urban areas and from public consciousness, making it 'up to us to make sure that we can be seen and heard.' Each performance becomes an assertion that Indigenous culture is not a historical artefact but living practice, not primitive tradition but a sophisticated knowledge system, not irrelevant to contemporary life but essential for addressing current challenges around sustainability, community building, and relationship with place.

She is grateful to build on the work of many women before her. The women who shaped this understanding of cultural responsibility span multiple generations. Aunty Winnifred (Winnie) Quagliotti, active in the 1970s and 1980s, demonstrated how cultural activism could work within mainstream institutions whilst maintaining Indigenous values. Her funeral, covered on all Australian television channels, marked her as 'probably the last one that was truly proper,' representing a generation of Indigenous leaders who had lived through the worst periods of government assimilation policies. Aunty Vicki Nicholson Brown provided different modelling through an approach that emphasised engagement. Mandy notes that she had 'just always been present and working in terms of cultural heritage, and she's just always been this beacon,' building trust and knowledge through decades of steady engagement.

Mandy also speaks of the deep influence of Borate (Annie), Berak's sister, who experienced the full force of policies designed to eliminate Indigenous culture. 'She lived through that time where her identity was told to stop and take on another one, spirituality as well,' says Mandy, yet despite these pressures she maintained connections that enabled cultural transmission across disrupted generations.

These women, along with Mandy's German mother, who survived her own forms of cultural destruction and abuse, demonstrated different strategies for maintaining strength, leading to Mandy's observation that 'the women have really kind of been the constant. That invader mentality has beaten those men down so far that they get lost in this world that was created around them, where women have always been there as nurturers and protectors of their babies.'

The strength of Aboriginal women in maintaining culture through the worst periods of government assimilationist policies enabled the cultural revival that benefits contemporary communities, including Mandy's own daughters. 'They are adults now, and they've become the mentors for the young girls that are coming through. If [the girls] decide to follow the wrong path, they can come back. They have the cultural competence to be able to speak up about who they are…freely, and they can reply to all these racist comments thrown at them, not with anger or swearing but speaking with knowledge.'

When asked about urban sustainability, Mandy speaks about how sustainability in an urban environment means 'rebalancing the ecosystem with cultural practice.' This integration challenges Western assumptions that separate human culture from natural systems, proposing instead that ecological health requires cultural health. She advocates for 'taking it back to all the old ways of cultural burning and sustaining the environment and putting balance back into the environment.' Cultural burning—the systematic use of low-intensity fires to maintain ecosystem health—represents an integrated approach to environmental management that emerged from tens of thousands of years of careful observation, requiring not just technical knowledge but cultural frameworks that understand landscapes as relationships rather than resources. It requires constant navigation between different ways of knowing, leading Mandy to describe her work as 'speaking for those that can no longer speak, or those leaders that we stand on the shoulders of, because it's not just my culture and my language—it's ours, and we just need to continue it through as strong and loudly as we can because we don't want to have to ever fall asleep again.'

As our conversation draws towards conclusion, Mandy's vision of cultural continuity takes on both urgency and playfulness. We ask her to imagine a final performance before global collapse, and she initially proposes a defiant 'mic drop' moment—a dance that tells the world, 'You effed this up, and you had all the answers and solutions right in front of you, but you did not take them on board.' But she is also drawn to the gentler message of the mother tree dance, which emphasises 'cultural survival and renewal and looking after each other and making sure the next generations are okay—not being selfish, but sharing, and just standing strong when a storm hits,' balancing accountability with hope, acknowledging the failures that have led to crisis whilst also affirming the knowledge and practices that can support regeneration.

15

Sheela Patel (India): The Village in the City

Abstract Sheela Patel is the founding Director of the Society for Promotion of Area Resource Centres (SPARC), a Mumbai-based NGO supporting urban poor communities since 1984. Sheela is internationally recognised for seeking urgent attention to the issues of urban poverty, housing, and infrastructure onto the radar of governments, bilateral and international agencies, foundations, and other organizations.

Keywords Informal settlements • Urban resilience • Social sustainability • Capacity building

5 December 2023, in Person at Dubai COP28; Updated December 2025

Rather than seeing informal settlements as problems to be solved, Sheela Patel sees them as 'places of innovation' where people create solutions to problems that formal systems have not addressed. Instead of seeing poor people as lacking capacity, she sees them as possessing knowledge and capabilities that development programmes consistently underestimate.

C. Oke, J. Verbeeck, *City Leaders for Sustainable Futures*,
https://doi.org/10.1007/978-981-95-9544-0_15

Fig. 15.1 Sheela Patel

Her work demonstrates that the most important knowledge often flows from communities to institutions, rather than the reverse. This perspective becomes particularly important when considering climate change and urban resilience. The communities she works with have been adapting to environmental stress, economic uncertainty, and social exclusion for generations, developing what she calls 'frontline' capabilities for surviving and thriving under difficult conditions (Fig. 15.1).

Speaking with the authority of 50 years spent bridging worlds—between formal institutions and informal settlements, between global networks and neighbourhood committees, and between the abstract language of development policy and the urgent realities of families seeking secure tenure—Sheela Patel settles into our conversation with assured confidence of someone with deep insights from working in the margins. Our interview in Dubai during COP28 breathes the decades spent translating between different ways of knowing. The first thing one would probably notice is how Sheela, always with humour and modesty, makes you believe that change really is around the corner. She does not need much of a nudge to share personal insights. Her professional and personal journey are deeply intertwined.

'Rather than mentioning titles, I want to share that I'm not an expert at anything. My skill sets lie in being able to encourage large cohorts of community leaders to start sharpening and defining their articulation of what they need, what they want, and what they can do or are doing themselves. And to keep women at the centre of that process. Not excluding everybody else.... But by starting with women, it becomes a growing cycle of inclusion for larger volumes. That's one. The second thing is the power of fabrication, which has always fascinated me. Creating systems and creating rituals and practices that are simple but that can be adopted by anybody and everybody. Those are crucibles in which you refine and sharpen strategies and that you then throw out for others to pick up, to learn from, to change. And to develop a confidence and capacity to feel that you don't need to own anything. To let your work be spun around and make it do different things. That is how real scaling works. Scaling doesn't work out of a cookie-cutter approach. It comes out of giving people a sense of empathy. To encourage them to take something that inspires them and twist it into something for themselves. When you give something, it goes around and comes back: an enrichment of what you started. So if you don't have an ego about protecting your baby, you say, "Okay, now have a grown up version of that." So I think that's my skill set—not to mention I am like a sponge.'

Rather than a Dubai conference room, Sheela would take us to informal markets that exist in every city: those spaces where the character of urban life shows itself most clearly, where the relationships between rural and urban, between different classes, and between traditional ways and contemporary pressures play out in daily transactions.

'I like informal markets. They show you the character of a city,' she explains. The markets she describes exist everywhere she travels, from Bangkok to Mumbai to cities across Africa, with each carrying the specific cultural DNA of its place. She recalls how some 30 years ago in Bangkok, 'both rich and poor people went to the wet markets, because that's where you got the freshest food. Apart from the commercial character, you built up long social relationships with the people in the store that created community across economic differences. The rich came in the morning to buy the best food, whereas by the end of the day, poorer communities would come and benefit from reduced prices as vendors needed to clear their stock.'

Over time, these markets also started to display global pressures that were reshaping cities everywhere. Sheela notes that when Asian supermarkets began competing by offering convenience and sanitation, 'the rich started going there. As a result the wet markets have stopped flourishing. They could not sell their best produce at the price line. The competition also caused the poor people's nutrition to be reduced.' This pattern—the disruption of systems that serve everyone to systems that serve only those with money—illuminates broader challenges, where supermarkets become microcosms of gentrification, demonstrating how the logic of efficiency and modernisation can destroy the very social relationships and affordability that made cities work for everyone.

Her own relationship with markets traces back to the childhood education she received from her grandmother. She vividly recalls one day, about age 10, going to the market and coming home with a so-called non-Indian chicken—imported poultry that looked impressively large but lacked the quality her grandmother could almost instinctively detect. 'From the look of it and the smell of it, you should know what's fresh and what's not fresh, what is local and what is not local. Different sorts of fruits, vegetables, fish, meat, spices, everything is a different lesson,' Sheela recalls. Mastering these distinctions meant that 'the food that came out of your kitchen came out of an appreciation of what you wanted to purchase.' The simple story reveals something about her approach to development work: the most sophisticated knowledge often appears as common sense to people embedded in place and community who have learned to read the world carefully (Fig. 15.2).

The object Sheela chooses to represent her journey carries the complexity of intergenerational transmission that shapes all her work. Rather than something from her professional achievements, she has chosen her granddaughter's paraphernalia photos: images that capture her own nine-year-old granddaughter's remarkable ability to understand and articulate

Fig. 15.2 An illustration of a market, the place Sheela would prefer to conduct this interview

environmental issues. Living in Vietnam and attending what Sheela's husband calls 'a hippie school' with teachers from around the world, her granddaughter has grown up with a climate consciousness as natural as breathing. Broadly smiling, Sheela shares a story of her granddaughter being grateful about having her as inspiration and how in a classroom discussion about climate she gave her classmates 'a big lecture' about sustainability.

Listening to Sheela, you wonder how this activism trickles down across generations. It feels like acceptance when she shares how her daughter 'competed with my work obsession' and found her mother's constant travel exhausting, whilst her granddaughter embraces this legacy with enthusiasm, offering to make videos for Sheela's Roof Over Our Heads campaign. Sheela has allowed her family to put limits on her activism. 'My family really contributes to my balance. I get a kick in my ass if I don't do all the things I'm supposed to do with my family. Every year, on the first of January, when we have a new calendar, I am given no-fly moments for all our festivities and family time. I think that my own ability to be sane comes out of that balance.'

Her smile is disarming when she evokes how her professional quest and family life are intertwined, but she is aware that generational change also comes with responsibility. 'We are taught that, you know, we're doing this for our grandchildren, but being an actual grandmother makes the work very real in ways that abstract intergenerational justice cannot.'

The transformation from conventional development thinking to her current approach began accidentally, through a job she took primarily to avoid 'being domesticated' as a young married woman. Sheela's first job was as a counsellor in a community centre in Mumbai. Working in India's second largest city, she encountered families who had migrated from rural areas because 'the ecosystem of their households were destroyed.' She says, 'Men came to work in the city and the women followed them, and these families had no place to stay. So they stayed on the pavements. There was this one woman on the street in a tent-like structure, her house literally this much (depicting with her arms a space not much wider than one or two arms wide from the body on each side). Five people live in this much of "a house." The woman had to steal, or buy, five, six buckets of water to cover their basic needs.'

When Sheela offered help, the woman's response was devastating: 'I can wash my face and rinse my hands, but I can't have a bath unless I go and work as a domestic servant in somebody else's house. Often I only have three buckets of water, so what is left for cooking then?' Seeing how this woman was so oppressed by receiving advice from 'some educated woman' was a transformative moment. She understood the impossibility of the advice but also where development actually works or fails and that it requires listening rather than prescribing and understanding actual constraints rather than imposing ideal solutions.

'These women don't have time to listen to esoteric advice. They see what you're doing and if it works, they do it too. So, how do you create conditions by which you give a few women an opportunity to try something new and say, in return, ask them to tell 500 people?'

The strategies that emerged consistently revealed how simple policy changes could transform lives when driven by actual need rather than abstract principle. Her work in Dharavi, the Mumbai settlement that houses half a million people in what outsiders see as an undifferentiated slum, demonstrates how community knowledge can transform urban planning. Her team provided residents with materials to map their own neighbourhoods. The results revealed '98 settlements which had borders which had names and had different forms of associations' within what planners had been treating as a single unit, whereas the community-produced map showed so many businesses and such complex social organisation that formal planning had never recognised. This work led to a book called *Re-Dharavi*.

'In [the book] we say that when you go to redevelop informal settlements as architects, designers, and planners, you should not treat it like a greenfield, ignoring everything and just rebuilding it. No, you have to look at the culture and the physical locations and the organisations that are there and develop an alternative for them. In most large cities, slums are huge. They are like your European towns: 500,000–800,000 people, structures, or even much more. Look at Mumbai. Of the 22 million people, 56% live in slums. Those are not small numbers. So how do you go about developing these townships or towns within cities and the character, identities, boundaries they have? It's not just some amorphous place. So now we have within Slum Dwellers International the capacity to tell cities and to governments that if you want to redesign or redevelop a large, informal township, you have to do it this way, not that way. You can't just take a ruler and just draw lines.'

The insight that informal settlements represent sophisticated urban organisation, rather than lack of organisation, has influenced policy discussions globally. Sheela emphasises that this seldom-recognised knowledge comes from residents themselves, not from external research. The international work that has consumed much of the last two decades has emerged from recognising that Mumbai's challenges exist everywhere and that innovations developed in one place can inspire adaptation elsewhere through the creation of the Society for the Promotion of Area Resource Centres (SPARC), the co-founding of Slum Dwellers International, and her role as global ambassador for Race to Resilience.

Her current work with Roof Over Our Heads represents an attempt to scale this enabling approach through technology whilst maintaining community control. Rather than creating another development programme, the initiative provides tools that enable women to assess and improve their housing conditions themselves, documenting their own progress and sharing knowledge with others facing similar challenges.

The broader vision that emerges from Sheela's work challenges fundamental assumptions about urban development, poverty, and expertise. Climate funding and resilience planning consistently overlook existing capabilities, focusing instead on technological solutions that may work in ideal conditions but fail when implemented in complex social and economic contexts. The COVID-19 pandemic provided a striking illustration of how, whilst formal systems struggled to respond to rapidly changing conditions, informal networks demonstrated remarkable capacity for mutual aid and community support.

'Do you need somebody to do a PhD or master's degree to contribute fully to their society in Africa? I don't think so,' she observes, challenging assumptions about formal education that often exclude community knowledge. Her work shows how practical wisdom and lived experience often provide better preparation for addressing complex challenges than academic credentials alone.

Looking towards the future, Sheela's vision combines ambitious goals with practical incrementalism. Though mentioning deep respect for leaders like Gordon Brown and Christiana Figueres, her 'dream team' for addressing global challenges mostly centres on 'next-generation leaders and voices from diverse backgrounds'—a team that brings lasting change through collaboration across different types of knowledge and capability, demonstrating that transformation requires multiple approaches working together, rather than any single intervention dominating the process.

16

Marvin Rees (UK): Finding Significance in Insignificance

Abstract Lord Marvin Rees, Baron Rees of Easton OBE, served as Mayor of Bristol from 2016 to 2024. He currently serves as member of the House of Lords. Marvin has received many awards recognising his approach to city leadership and impactful governance.

Keywords Transformative leadership • Equity • Collaborative urban governance • Political capacity

4 March 2024, Online from Bristol, England

Approaching one of his final interviews as mayor of a city he has led for nearly eight years, Marvin Rees settles into our conversation with the reflective quality of someone preparing to close one chapter whilst opening another (Fig. 16.1).

The late-afternoon light filtering through his Bristol office illuminates not just the end of a mayoral tenure but the culmination of a journey that began with an 18-year-old's moment of profound recognition on a distant Arctic mountain in Spitsbergen, Svalbard, where he discovered what

C. Oke, J. Verbeeck, *City Leaders for Sustainable Futures*,
https://doi.org/10.1007/978-981-95-9544-0_16

Fig. 16.1 Marvin Rees

he describes as 'my significance in my insignificance.' Above the Arctic Circle, surrounded by endless mountains and glacier crevasses, he experienced the paradox of feeling simultaneously fragile and profoundly alive.

It already provides an indication of where he would meet us: a summit within that glacial mountain area, where his breath was taken away when he first came upon it as a teenager with the British Exploring Society. After weeks of preparation and teamwork with three companions, they climbed towards the summit, emerging at midnight sun onto a vista of 'thousands of mountains' stretching endlessly in every direction. 'I thought if my body went down there, no one would ever see me again,' he recalls. Yet rather than feeling diminished by this immensity, something profound shifted: 'I was acutely aware that I was alive, and I existed, and it's been a profound, profound moment.' This experience of simultaneous fragility and significance would become foundational. 'I was aware of the smallness and the fragility of my life, my physical body, but I was alive, and I knew I was alive. But that contradiction was quite profound.'

He continues, 'Sometimes I just like being by myself, also in the middle of busy places. To watch all the lives go by, a bit like in *The Matrix*, where Ray is walking through and suddenly the movie makes you acutely aware of the lives of all the people around. There is this place in Bristol I used to go to a lot when I was in my teenage years, the sea wall on the far side of Clifton Downs. I used to run up there from Easton, in our inner city, to get the feeling of being on the edge of the city. Then I just sat looking over the sea wall. Or on Brandon Hill, where you can find Cabot Tower. On the top of that tower, there are these brass plates. And on them they've got the names of cities around the world with an arrow pointing in which direction they are and the number of miles. I used to love going up by myself and imagining how by switching direction I would face Paris, Madrid, or New York, and all these lives over there. It took me out of the confines of the city. I don't like feeling trapped.'

For someone whose early life had provided few external confirmations of worth or possibility, this moment offered internal recognition that transcended social categories, creating what would become his approach to leadership, which holds both humility and authority. The journey to Spitsbergen was suggested by his schoolteacher, Mr. Jennings, representing what Marvin identifies as a pattern of proactive support. 'Someone came and made an intervention in my life and put me in touch with an opportunity they thought would work for me,' he explains.

The 18-year-old who climbed that mountain was carrying multiple forms of disadvantage. 'I grew up a mixed-race child of a single white woman. We were poor. We didn't have much money and were on benefits. It's only over the last two years I've realised that actually a lot of my political motivation is not even just about my experiences but those of my mum.'

As he was born in 1972, his very existence challenged social norms, where 'some health workers were telling my mum that if she was a good woman she'd have me aborted. Then once I was born, she was told if she were a good woman, she would give me up for adoption.' These early experiences of being categorised as illegitimate, mixed race, and poor created a paradoxical internal landscape where 'there's nothing in my life that tells me I exist or that I have any significance, apart from the level of my family,' yet simultaneously, 'I always believed in God, whatever anyone means by that. I had a sense of light and meaning, a sense of worth that transcended what people thought about me.' Rather than resolving the tension, he learned to inhabit it productively. 'This again is significance in my insignificance. Whether my smallness in the mountain with nothing to validate me, or smallness in the economy, politics, society with nothing to validate me—I found an awareness that meant my existence mattered in the face of these giants. It's not that I was positive in insistence here. It's that I could not recognise that I had a glimpse of my significance—and having [once] glimpsed it, I knew it was there, even when I couldn't feel it or see it.' (Fig. 16.2)

For Marvin, global awareness and local responsibility have always been intertwined. What symbolises his journey into sustainability thinking are accumulating experiences from his work with Tearfund, a development agency where he witnessed firsthand 'the precariousness of life for some people' in West Africa. He travelled to Guinea Bissau, rural Ghana, Liberia after its civil war, and the Philippines in the aftermath of Mount Pinatubo's eruption, including time in Quezon City to spend time with Samaritana, a charity working with women caught up in the sex industry. He began to understand what we call a 'just transition': how economic frameworks and climate impacts interact to determine human resilience. 'The fragility of people's lives in that sense was quite a profound intervention for me and my sense of wider social responsibility,' he reflects,

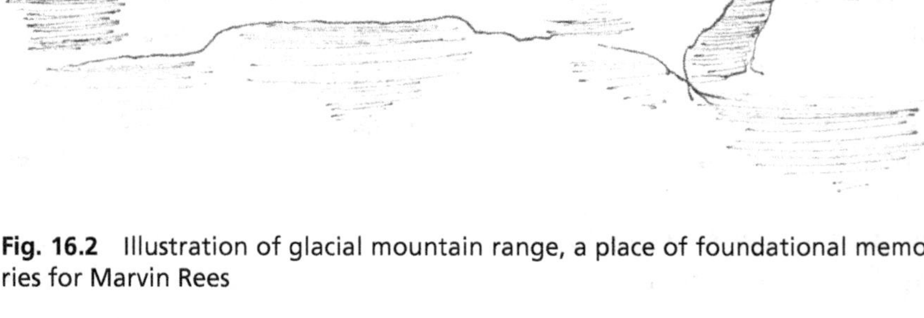

Fig. 16.2 Illustration of glacial mountain range, a place of foundational memories for Marvin Rees

describing how these experiences demonstrated 'low levels of resilience' where 'one subtle change in the environment could unlock a whole sequence of events in people's lives that result in conflict and/or poverty, death and/or migration.'

His involvement in the Jubilee 2000 debt cancellation campaign deepened this systems perspective, revealing how 'the financial framework within which people lived within their countries and in the financial framework around their countries had really undermined whatever post-colonial structures that were left in place that gave them the ability to cope with any challenge or catastrophe that comes along.'

The continued support during his journey spans from local mentors to global figures. Locally, businessman Andy Street provided crucial early support, believing in Marvin's potential when Marvin himself felt that he lacked credibility as a mayoral candidate: 'I was on stage with three white men who were late 60s, early 70s, a retired GP, a semi-retired architect, and a retired accountant. And then there's this young Black guy, 6 ft with a shaved head, on the row. I didn't think anyone would be looking at the stage and thinking, "He looks like a mayor."' Bristol had never had Black political leadership. When elected, Marvin became the first person of Black African heritage to be elected mayor of any European city. Similarly, his MP, Kerry McCarthy, took him seriously at a moment when self-doubt threatened to overwhelm possibility.

Historically, Martin Luther King Jr. left a mark. It was not so much the 'I Have a Dream' speech that resonated with him but 'Beyond Vietnam: A Time to Break Silence,' a speech King delivered in 1967. 'That one shows the unpalatable King, the one they thought was dangerous and should be stopped, that the one who went and stood with sanitation workers with a cause that went beyond morality to the economic and political, America and the world.'

His approach to mayoral leadership reflects lessons learnt from all these influences, particularly the recognition that effective change requires what he calls 'productive use of time' rather than political theatre. 'If I would draw a pie chart of the use of time mixed with important use of time, the back and forth of party politics is a very small sliver of that. If we actually want to get things done, the vast majority of that pie must be spent interacting with the government, government agencies, private sector builders and investors, the voluntary community sector, public sector partners, and criminal justice organisations.' He meets monthly with over 30 community development workers from across the city and biweekly with chief executives and heads of key institutions including police, universities, unions, and the private sector. Every six months, 'we have something called a city gathering where we bring around 350 to 400 people together from all backgrounds running organisations in the city, and it's like a city village meeting.'

This strategy emerged from his background in public health and the idea of 'wider determinants of health': that there is no direct connection between health services and health outcomes. 'There's no straight line between pulling a political lever here and getting a public outcome there. What citizens receive is not the result of any single organisation's decisions or non-decisions. Citizens are at the intersection of the whole complex range of decisions or non-decisions made by all institutions that shape a place.' This led to the One City Approach, which brings together leaders from across sectors in the Bristol One City Plan. 'Within two months of getting elected, I convened around 70 leaders of the key organisations across the key sectors. I stressed our interdependencies: population health impacting on economic resilience, health depending on housing, and the future workforce depending on today's children and young people. I shared that 'between us we have about a six and a half billion pound footprint in Bristol, and we employ 70,000 people."'

The approach proves particularly important given Bristol's challenges: population growing from 460,000 when Marvin was elected to a projected 550,000 by 2050, 22,000 people on waiting lists for housing, extreme educational inequality where 'in Clifton, almost 100% of teenagers go on to higher education; in Hartcliffe it's about 11%.'

'I think the density of urban life offers profound opportunities and challenges. Living in such proximity to each other represents quite a profound shift in a very small amount of time in our evolution as beings.' Cities he sees as 'almost post-national entities' where 'Bristol is 180 countries of origin, 92 languages, 45 religions. They operate and they need to be allowed to operate in many ways beyond national boundaries. Take Bristol's music scene, for example. It is not just a product of one culture but the interaction of all different cultures coming together and creating something new.'

This cultural creativity matters politically, creating a potential antidote to rising isolationist, nationalistic authoritarianism. 'If done right, because of the proximity and need for global connectivity, cities have a head start, though the success is not guaranteed. If cities are starved of resources and have inequality and deprivation, global connectivity (and migration) could exacerbate tensions. Populists, racists, misogynists feed on communities who have been worn down, who have a weak hold on hope.' The metaphor he uses draws from Tolkien: 'What if we would get as much power as possible into the hands of mayors to keep it dispersed and safe? Almost like *Lord of the Rings*. You don't give the ring to the king, you know, to look after. You give the ring to the hobbit.'

His personal approach to sustaining this work combines acknowledgement of anger with strategic management of how a Black man allows that anger to be expressed. When Extinction Rebellion activists told him he did not 'seem to be angry enough' about the climate crisis, his response was immediate: 'How do you know when I'm angry? First off, you don't see how I must manage myself. Certainly public display of anger is a white privilege. If I start to be angry publicly, I don't go well. This is the stuff that gets us killed, right?' This recognition of how racial dynamics shape leadership possibilities runs through his understanding of the mayoral role, where 'I was apparently the first Black person to be elected mayor of any major European city.'

Looking towards the future, Marvin's vision for his own children displays his understanding of how to prepare young people for leadership in an unjust world: 'I tell my children I want them to have three assets: I want them to be intellectually strong, physically fit, and have good character. And if they build those three assets in their lives, they'll have agency and a degree of immunity to the negative decisions of others.' He adds two more: spiritual awareness and having a purpose. 'I told them that as brown people, the world isn't going to treat them the same as white people. It's not going to be fair. I'm just being real. This is why you need to build your assets. If you build your assets, you will build your resilience in the face of other people's decisions compared to your own.'

17

Aromar Revi (India): The Systems Weaver

Abstract Aromar Revi is founding Director and Vice Chancellor of the Indian Institute for Human Settlements (IIHS), the world's first urban-focused interdisciplinary university focused on sustainable urban transformation. A globally ranked scholar in climate adaptation and urban sustainability, he has served as Coordinating Lead Author for multiple IPCC reports and co-chairs the UN Sustainable Development Solutions Network.

Keywords Systems thinking • Climate adaptation • Urban resilience • Sustainable urbanisation

12 March 2024, Online from Bangalore, India; Updated December 2025

Having just returned from three days at the Vatican 'trying to make sense of what the hell is happening across the world,' Aromar Revi appears on screen, as usual taking his time to settle into the conversation. Behind him on the wall hangs a photograph of Central Park. It provokes an

C. Oke, J. Verbeeck, *City Leaders for Sustainable Futures*,
https://doi.org/10.1007/978-981-95-9544-0_17

invitation for an imaginary walk, though a cautionary one, as he explains: 'Having this picture behind me is a reminder of how bad it can get. You could build the richest, most productive metropolis of the nineteenth century, and then you forget to have green spaces inside. How bad can it get?' (Fig. 17.1)

Over the years, Aromar has often referred to that bigger picture about how even our greatest achievements can embody our greatest failures and how the work of building sustainable cities requires constant vigilance against the seductive logic of high modernism that prioritises efficiency over humanity, growth over well-being, and technical solutions over systemic understanding. The Central Park example serves as more than a

Fig. 17.1 Aromar Revi

Fig. 17.2 An illustration of Central Park, New York City, a constant reminder for Aromar Revi

reminder: it was created only after Manhattan's development pattern had eliminated existing green space, and it required massive public investment to retrofit what could have been preserved through different development patterns (Fig. 17.2).

He resists the premise entirely when asked what would be the ideal place to have this conversation. 'It's a tough question because I can pick up that talk literally in every continent and in multiple places. It is not about the celebration of place, rather about the maybe 100 different cases that could trigger change. The places that I've actually spent time in and done some of the most interesting work are places that are still very rural,' he explains. 'That's why territoriality is such an important part of our experience.' He does not see cities as discrete entities but as nodes in larger social-ecological systems that extend across landscapes and through time.

This systems perspective was forged through early encounters with foundational texts. Reading the Club of Rome's *The Limits to Growth*[1] in 1979, as a young person with training in technology and the hard sciences but interests spanning the humanities and social sciences, opened up 'a whole range of new areas of inquiry.' The book's insight that 'you could actually change and shift dramatically different things in the world if you really understood how systems actually worked and functioned' became a guiding principle.

This early exposure led through encounters that seem almost mythical: working on a background paper for the Brundtland Commission in 1985, presenting at Harvard, arriving at Dartmouth College to find Dana Meadows's office adorned with the famous quote attributed to Margaret Mead about how 'a small group of thoughtful, committed citizens can change the world,' and eventually being invited to join the Balaton Group, named for the Hungarian lake where they have gathered each September for now more than three decades.

The Balaton Group became a crucible for developing intellectual frameworks that would later inform everything from IPCC assessments to the Sustainable Development Goals. 'For almost 20-odd years, from 1988 onwards, I worked with them; some very interesting and remarkable people as part of that. The key question for us was, "What is the future of humanity?"'

This dual life—spending eight or nine months each year working on rural development projects across India and then gathering with global systems scientists to model planetary futures—created a unique perspective. The group helped rebuild the World3 model that had informed the original Limits to Growth study, producing World4 just in time for the Rio Earth Summit in 1992, with sobering results: 'What we're seeing in today's world, even though we didn't have the jargon for it then, was already very clear and obvious in the model's crash-and-burn scenarios.'

Dana Meadows—a world-renowned scientist who left her tenured position to farm in New Hampshire—became a profound influence on

[1] Meadows, D. H., Meadows, D. H., Randers, J., & Behrens III, W. W. (1972). The limits to growth: A report to the Club of Rome. https://www.library.dartmouth.edu/digital/digital-collections/limits-growth

Aromar's understanding of how change happens. She embodied the integration of thought and practice that would become central to his own work. 'She said: "Look, if I can't farm, if I can't do the basic things that need to get done in life, then I'm not able to practice what I'm talking about, I can't teach, I can't research, I can't be useful."' Her method of teaching—working in the potato fields whilst discussing complex mathematics—provided a template for the kind of integrated education Aromar would later seek to create. The combination of rigorous analysis with hands-on practice and a global perspective with local engagement became hallmarks of effective sustainability education. The transition from global modelling back to grounded urban work came through India's rapid transformation in the 1990s, as liberalisation created new challenges requiring going back 250 years in time to understand how urban systems had functioned before colonial disruption.

A pivotal project in Lucknow, a city of then 2 million people (now closer to 4.5 million), revealed sophisticated local approaches. For example, graveyards and 'places of rest'—sacred spaces that could not be built upon and served multiple functions simultaneously—occupied the lowest areas of each neighbourhood, where water would naturally flow into open wells that provided natural filtration, and were year-round green spaces embodying the Islamic vision of paradise. 'These guys actually got it right,' Aromar reflects. 'These are cultural spaces with a long tradition, serving communities that have been there for a few hundred years. The delivery of very basic things like water, sanitation, education, healthcare—you need to put in that context.'

Later work on sustainable urban futures initiatives in Vancouver, Shanghai, the Netherlands, and Goa reinforced his understanding that sustainable urbanism requires approaches that are simultaneously globally informed and locally grounded. Each context presented different opportunities: Vancouver's environmental consciousness provided space for experimentation but struggled with housing affordability; Shanghai's rapid growth created possibilities for large-scale innovation but raised questions about social inclusion; Dutch approaches to managing water and density offered technical solutions requiring adaptation to different cultural contexts; Goa's development pressures demanded approaches that could balance economic growth with environmental protection and social equity.

Collaboration with colleagues like David Satterthwaite and Debra Roberts brought him into IPCC assessments that helped establish the importance of cities in global climate action and to reframe them as 'places of opportunity and not a real mess.' Their work helped international climate policy look beyond national governments and demonstrate the importance of urban governance for emissions reduction, adaptation, and resilience.

The reframing extended into negotiations that created the Sustainable Development Goals. A weekend phone call and an instantly written briefing paper for the UN High-Level Panel during the SDG drafting process would lead to SDG 11 (on sustainable cities). The roller coaster continued, mobilising mayors from around the world, even crashing Bloomberg's Christmas party in New York to make the case for urban inclusion. Yet for all his involvement in global processes, Aromar remains grounded in understanding that real change happens through patient implementation.

His current efforts focus on building the Indian Institute for Human Settlements (IIHS) as a space where the kind of integrated thinking he learnt in the Balaton Group can be embedded in new generations. IIHS represents attempts to institutionalise transdisciplinary approaches. Its campus serves as a living laboratory, bringing together social, ecological, and built environments, whilst the curriculum integrates architecture and planning with ecology and economics, social science with engineering, and traditional knowledge with contemporary technologies.

Implementation challenges are illustrated through IIHS's work on sanitation systems in Tamil Nadu, a state in southern India. There his team has built what he describes as 'the largest SDG 6 [on Clean Water and Sanitation] implementation programme in the world.' The technical innovation—creating hybrid systems combining traditional practices with modern infrastructure to serve 20 million people—is impressive, but the deeper story is about social transformation, addressing the reality that 'even in 2024, 95% of the people who clean toilets are people who are called Dalits—historically untouchables.' The project seeks to change not just infrastructure but social relations, addressing deep inequalities that shape how cities function, who benefits from urban development, and who bears the costs of urban environmental problems.

'In the end, the larger story for us is one of a story of social transformation,' he says. This attention to inequality informs his complex relationship with dominant narratives about urban opportunity. Whilst these narratives are instrumental in promoting cities as spaces of possibility, he also insists on acknowledging what he calls 'the dark side of cities.' 'In the Western context, since the time of the Greeks and the Romans, we have known that cities concentrated poverty and also slavery. Pre-industrial cities have shown tremendous productivity, building on slavery, indentured labour, and the systematic exclusion of women and other subordinated groups.... We are often still reproducing this in the current context, but we don't talk enough about it.' The COVID-19 pandemic accelerated these dynamics, concentrating wealth whilst impoverishing everyone else. 'Poverty is different from inequality,' Aromar notes. 'If you look at COVID, pretty much every individual in the world, including billionaires, fell ill, except that the wealthiest people got richer because their wealth and capital earned interest, while people were dying.'

This analysis leads to profound questions about the future of urbanisation itself. Drawing on an understanding of longer historical cycles, Aromar wonders whether planetary urbanisation represents an inevitable future or whether different patterns might emerge. In a world of nuclear weapons designed as 'city killers,' the concentration of population creates unprecedented vulnerabilities. Yet rather than leading to pessimism, this understanding creates urgency for cultural transformation. 'Can we build a global or regional or local culture of sustainability?' Aromar asks. 'The answer is we don't know. We have utopias, we have intentional communities in some places, but we don't have adequate imaginaries for the twenty-first century.'

The search for new imaginaries leads him to reflect on examples like Auroville, the international community near Chennai that has been 'building the city for the future' since 1968. Despite remarkable achievements—for example, growing 2 million trees in once-barren land and creating innovative approaches to sustainable living—the community has remained at 5000 people rather than growing to the 100,000 that might represent a viable alternative. 'There are many experiments like this in many parts of the world,' Aromar observes. 'Either the big ones are constrained because of scale and structure, or the small ones are unable to build up.'

His role, as he sees it, is to create conditions where these different levels of change can reinforce each other; where insights of systems science can inform practical projects; where local innovations can scale to influence global frameworks; and where patient education and institution building can prepare new generations.

IIHS University, of which Aromar is the founding Vice Chancellor, is the first urban-focused interdisciplinary university in the world and constitutes an attempt to institutionalise this integrated approach. Its campus demonstrates these principles through design that integrates built, natural, and digital environments, creates spaces for both individual reflection and community gathering, and incorporates traditional building techniques with contemporary technologies.

'It is often about time, is it not?' says Aromar. 'Cities are very fragile things, you know, 5000 years of history, and we still have no clue how to actually manage them. There is a big risk that we might see a scale of conflict in this century that we've never seen before. Take the current global food trade system, for example, that should help to keep the world alive. But half a billion people cannot produce or buy enough food for themselves. There's a huge and growing imbalance in the global system. And cities can exacerbate that. Especially if we over-urbanise or urbanise too fast, or even build [the] mega-urban regions that are popping up across Asia. Nobody seems to wonder where their food or water will come from.'

'This century will hold something like three quarters of a trillion person-years of living experience, based on, let's say, 10 billion people, each with a mean life expectancy of 70 or more. That is more than all the human lifetimes since the beginning of history. The risk of losing a large share of much human potential due to the climate or food crises is huge, which is where sustainable cities will play a pivotal role, with their unprecedented potential, opportunity to connect, to accelerate transformational change.'

'That is why education and dialogue are absolutely critical. But I feel the world is still organised around nineteenth-century and twentieth-century principles. Applying those practices and knowledge to urban planning in a twenty-first-century world that's overshooting planetary boundaries is a huge mismatch. That's why imagining possible futures is so important—and sharing with each other experiences of places that people can look to that offer glimpses of a different future.'

This illuminates why Aromar's work has focused on building educational and research institutions, rather than only pursuing more immediately visible forms of impact. He understands that the challenges facing urban civilisation over the next century require new forms of professional knowledge and institutional capacity that current systems cannot provide. The three-quarters of a trillion person-years that will be lived in cities represent unprecedented opportunity for human learning and development, and for cultural innovation and technological advancement, to address challenges that have seemed intractable when approached through existing institutions and frameworks.

18

Debra Roberts (South Africa): Under the Acacia Tree

Abstract Professor Debra Roberts is Honorary Professor at the University of KwaZulu-Natal (School of Life Sciences), Professor Willem Schermerhorn Chair in Open Science from a Majority World Perspective at the University of Twente in the Netherlands, and between 1994 and 2024 head of Sustainable and Resilient City Initiatives Unit and Environmental Planning and Climate Protection Department in eThekwini Municipality (Durban, South Africa). Debra was Co-chair of Working Group II of the Intergovernmental Panel on Climate Change (IPCC) during the Sixth Assessment cycle and is a Co-ordinating Lead Author for the IPCC Special Report on Climate Change and Cities.

Keywords Urban resilience • Climate risk • Biodiversity • Adaptation strategies

C. Oke, J. Verbeeck, *City Leaders for Sustainable Futures*,
https://doi.org/10.1007/978-981-95-9544-0_18

7 December 2023, Online from Dubai, United Arab Emirates; Updated December 2025

The festival atmosphere of COP28's fourth day pulses through Dubai's air-conditioned pavilion maze, where Professor Debra Roberts moves with the pace of someone who has spent three decades translating between science, policy, and community action. Recently retired as Co-chair of Working Group II for the Intergovernmental Panel on Climate Change's (IPCC) Sixth Assessment cycle, we are keen to peek behind her trademark beaming smile and fierce intelligence that has driven boundary riding work between university laboratories and township meeting halls and between international climate negotiations and Durban's informal settlements. There's something both accomplished and humble about her presence—the quality of someone who remains fundamentally curious about the world and eager to listen, even after decades of being sought out for her expertise on urban climate resilience. The path that brought her to this moment began not in conference halls or academic institutions but under an African acacia tree in the dry-land savannas outside Bulawayo, Zimbabwe, where a young girl learned lessons about limits, relationships, and reciprocity that would shape her understanding of sustainability long before she encountered the word in any textbook (Fig. 18.1).

If Debra could choose our meeting place, she would transport us instantly to Bulawayo, Zimbabwe, where we would sit beneath just such an acacia tree. Not the Australian-stolen name, she clarifies with characteristic precision, but a true African acacia—the low-growing tree whose thick, black seed pods hang to the ground like iron bars, creating natural caves where curious children might hide away from the world. This choice speaks to something fundamental about Debra's understanding of sustainability and resilience: that the most profound insights emerge not from abstract theory but from intimate encounters with place.

Growing up in Marlands, a small suburb outside Bulawayo with few houses, dirt roads, no street lights, and no waste collection, Debra experienced what she now recognises as 'an extraordinary privilege that I fear kids today don't have.' Her entire childhood was spent roaming beyond the boundaries of the family property, engaging in 'unstructured play in

Fig. 18.1 Debra Roberts

the world' whilst her mother and younger sister visited family in more formalised parts of Bulawayo. She would stay home with her father, spending entire days reading under trees in the bush, finding in nature what she describes as 'a safe space' even when human systems had descended into violence during Zimbabwe's brutal civil war of the 1960s and 1970s.

This experience of nature as refuge shaped her understanding of what communities need to build resilience—not just technical infrastructure or policy frameworks but spaces and practices that provide emotional and spiritual sustenance. The child who found sanctuary under acacia trees would later advocate for urban green spaces not just as environmental amenities but as essential infrastructure for community well-being (Fig. 18.2).

The woman who taught her how to read the landscape was her mother, who had grown up on a farm and came equipped with knowledge that would prove formative. Rather than staying inside to discuss distant concepts, 'she would walk with us and point and demonstrate,' showing them which berries could be eaten and, importantly, which could not, teaching them to play with bones of wild animals as toys when there was no money for more sophisticated entertainment. But the deeper teaching was about relationship and restraint: observing wildflowers as they emerged after summer rains whilst learning that 'while one should observe and enjoy, one shouldn't in fact harvest unnecessarily.'

These lessons about reciprocity and restraint would later inform Debra's approach to ecosystem-based adaptation, in which she advocates for urban development strategies that work with, rather than against, natural systems. The traditional knowledge her mother shared—about which plants could be harvested and when, about how to use resources without depleting them, about the importance of leaving enough for other species and for future generations—provided a practical foundation for understanding sustainability that formal education often lacks.

Fig. 18.2 An illustration inspired by Debra Roberts' memories of sitting under trees as a child

The book that most powerfully shaped young Debra's worldview was Frank Herbert's *Dune*, which she would read in her hiding place beneath the acacia's heavy boughs. This choice demonstrates something essential about her approach to sustainability: the recognition that environmental challenges are simultaneously technical and spiritual, requiring both scientific understanding and deeper wisdom about how humans relate to the systems that sustain them. Herbert's vision of the book's desert planet Arrakis provided a template for understanding how human actions ripple through complex systems whilst also exploring 'the spirituality that's associated with a close association with a place, with a geography, with an ecosystem.' This early integration of scientific and spiritual understanding would prove crucial, teaching her 'to be resilient and how to bounce back' as well as 'to stick to my convictions regardless of the potential professional fallout.'

Letting her memories speak, Debra pictures a kaross (the word is Afrikaans with origins in the Khoi-San language), a blanket made from dassie (rock hyrax) skins sewn together with animal sinews using an acacia thorn. This particular kaross was crafted by a San man called Kleinbooi (so named because Western tongues could not manage the clicks of his Indigenous language) who had travelled with her grandfather across the southern African continent. The blanket, made from animal skins, embodied the kind of complete resource utilisation, where 'every element of the animal was used, so nothing was wasted,' that would influence her thinking about sustainability for decades to come. But the deeper teaching came through her mother's stories about how Kleinbooi harvested honey from ground-nesting bees, using smoke to calm them whilst taking honey for human use but leaving sufficient stores for the bees themselves and for the honeyguide bird that led him to the hive.

'It was that kind of approach: you had a right to harvest, but you also had to share and leave things for the producers of the resources and your allies,' Debra says. This represents not just technical but ethical sustainability grounded in recognition of interdependence and responsibility. The kaross embodies principles that would later inform her advocacy for circular economy approaches to urban development, where waste from one process becomes input for another and where human needs are met through systems that enhance, rather than degrade, ecological function.

The transition from these early experiences to professional environmental work came through encounters that exposed both the fragility of natural systems and the possibility of restoration. During Debra's second year at Natal University, a field trip to Mtunzini introduced her to Ian Garland, a farmer whose childhood paralleled her own but whose adult life confronted him with the consequences of ecological destruction. Garland had grown up swimming in the clear brooks of Twin Streams farm, surrounded by lush coastal forests, but by the time he inherited the property, much of the forest had been cleared for sugar cane, and the streams had silted up to mere centimetres deep. His response was to dedicate his life to restoration, creating an educational centre where schoolchildren and university students could witness ecological healing in action.

This encounter sparked Debra's recognition that restoration was possible, that human intervention could heal as well as harm, and, more importantly, that her childhood experiences with nature could become central to her professional life rather than remaining separate from it.

Her eyes light up when she starts talking about John C. Poynton (known to everyone as JC), the professor who most profoundly shaped her understanding of how to integrate diverse ways of knowing. She describes him as 'probably one of Africa's best herpetologists' but also 'someone who studied parapsychology, was a classical music composer, and was deeply interested in philosophy.' As one of the postgraduate students who benefitted from his extraordinary knowledge, Debra received an education that defied conventional disciplinary boundaries, discovering 'this very big vision of science' that embraced everything rather than limiting itself to test tubes and controlled experiments. When Poynton assigned her to investigate why frogs were disappearing in Durban, he was introducing her to what would become her life's work: urban conservation, even though others said, 'This is not where biodiversity happens. What are you doing?'

Poynton's requirement that she read philosophical works about conservation, particularly the writings of Arne Naess, introduced her to the concept of 'deep ecology' and the insight that would guide everything that followed: 'You as an individual cannot save the planet. But what you can do is choose a place and defend it.' This place-based approach to environmental action resonated deeply with someone whose understanding of ecology had been shaped by intimate knowledge of particular landscapes, providing a framework for translating broad environmental concerns into specific commitments that could be sustained over decades.

Durban would become her chosen place, creating what she describes as terrier-like determination: 'Once you've got a cause, once you've got a place, then I'm like a terrier. I won't let it go.' The move from academic research to municipal practice came during South Africa's transition to democracy, when 'there was a mood in the country that anything was possible.' After finding formal academia constraining, Debra joined local government with the conviction that she could use scientific knowledge to drive real-world change, only to discover that effective municipal work required far more than technical expertise. 'It's all about people. It's who knows who, who owes who, who knows where the bodies are buried...that determines how that institution really functions,' she says. This recognition led to her distinctive approach to sustainability work: prioritising human relationships and storytelling over purely technical solutions, with an understanding that 'at the end of the day, we simply are a primate and need to interact socially and engage and tell our stories so that we can move forward together.'

When Durban began working with the 100 Resilient Cities programme, community and stakeholder engagement took the process in unexpected directions that manifested profound insights about resilience and adaptation. Initially expecting to focus on water, biodiversity, and climate change issues, the consultation process highlighted the importance of 'informality and governance' to the city's resilience. This did not sit well with some members of city leadership, who felt that 'informality is not something we should prioritise, as that could influence how investors view Durban,' Debra recalls. But engaging with informal communities revealed that informal settlements represent 'almost an act of protest, saying, "This development path is not working for us. So, we're going to build a city within a city to better meet our needs."' Rather than problems to be solved, these communities embodied innovative approaches to urban living that formal planning could learn from.

One of the most powerful examples came during major floods in Durban, when a partnership between an informal community, the university, and local government meant that in one particularly vulnerable settlement, despite houses and property being destroyed, no lives were lost, simply because they were treated as 'a community that we could talk to' rather than a technical problem merely requiring engineering solutions. This experience demonstrated that effective adaptation requires combining technical knowledge with social knowledge and that resilience emerges from community capacity as much as from infrastructure investment.

The challenge of maintaining this approach whilst operating at multiple scales—local, national, and international—required what Debra describes as a 'scissor approach': working simultaneously both top-down and bottom-up, rather than choosing between them. 'I had to do innovative local work to create the credibility of being able to do and deliver. But I also had to have a top-down approach. I needed to create a context within which that innovative work had permission to take place and could be valued.' This led to her involvement in climate negotiations as a South African government representative, her role as lead author and later Co-chair with the IPCC, and her current work on the Special Report on Climate Change and Cities.

The personal cost of this boundary-riding work—operating simultaneously as scientist and practitioner, local implementer, and global influencer—cannot be understated. 'It's mindlessly hard work, and it's mindlessly unsexy. There are no ribbon cuttings, there's no accolades. It's just really hard work in the trenches,' Debra says. The work requires staying in uncomfortable spaces, persistently challenging systems that resist change, often working alone or with minimal support.

But the results of such persistence can be transformative, as demonstrated when a former mayor acknowledged Debra's three decades of environmental advocacy: '… over time, we realised you were right. Now we're learning from you and listening to what you have to say.' This acknowledgement represents something more than personal vindication—it demonstrates how sustained engagement can shift institutional culture.

The lessons Debra draws from this experience challenge much of the contemporary thinking about change-making and leadership development, emphasising the need for boundary riders who 'stay where the people are,' committing to place and relationship over personal advancement, 'it is always about the mission rather than the position.' She adds, 'You've got to stay in the room long enough to know each other as people, to generate respect and trust,' requiring what she calls 'a certain amount of bravery'—the willingness to ignore instructions, contradict superiors, and persist in the face of resistance whilst building the relationships that enable transformation to occur.

Looking toward the future, Debra sees the IPCC Special Report on Climate Change and Cities as an opportunity to create global permission and recognition for locally grounded climate action. 'It is going to create a global narrative of cities that will provide permission for innovative local action, because the IPCC's assessment will provide scientific evidence supporting the strategic importance of local-level work that will make it harder for national governments to ignore.'

Through her story, we see how childhood encounters with the African savanna can shape global climate science and how the patient work of relationship building in one city can influence international frameworks for urban resilience. Her journey from reading *Dune* under an acacia tree to co-chairing an IPCC Working Group illustrates the power of sustained commitment to place combined with strategic engagement across scales of influence. The acacia tree of her childhood memory continues to provide shelter and teaching, its deep roots and spreading canopy offering a model for the kind of grounded yet expansive approach to environmental leadership that our changing world desperately needs.

19

Maimunah Mohd Sharif (Malaysia): Rivers of Memory

Abstract Dr. Maimunah Mohd Sharif is a Malaysian urban planner with over 40 years of experience in sustainable development. She served as UN-Habitat Executive Director (2018–2024), the first Asian woman in this role. A former mayor of Kuala Lumpur and of Penang and the first woman President of Seberang Prai Municipal Council, Maimunah now serves as Property Advisor at PETRONAS.

Keywords SDG11;·Sustainable urbanisation • Gender leadership • Local government • Urban planning

30 January 2024, Online from Nairobi, Kenya

Maimunah Mohd Sharif settles into our conversation just days before retiring as Executive Director of UN-Habitat. the world's most important urban development agency. Her six years at UN-Habitat have coincided with unprecedented urbanisation, with more than half the world's population now living in cities and that proportion expected to reach two-thirds by 2050 (Fig. 19.1).

C. Oke, J. Verbeeck, *City Leaders for Sustainable Futures*,
https://doi.org/10.1007/978-981-95-9544-0_19

Fig. 19.1 Maimunah Mohd Sharif

From her office in Nairobi, the complexity of contemporary African urbanisation is visible: the gleaming towers of the business district rising alongside informal settlements where millions have created homes on land they do no't officially own, where innovative entrepreneurship flourishes despite lacking formal support, and where traditional community structures adapt to modern urban conditions whilst maintaining essential social functions.

She does no't have to think long when asked where she would have taken us to record this interview. Not in any of the 90 countries she has visited in her UN role, nor to the gleaming conference halls where global urban policy is negotiated, but on a simple bench in her home village of Kampung Sungai Besar Dua in Negeri Sembilan, Malaysia, two hours from Kuala Lumpur. When she closes her eyes, she pictures sitting on the bench at the convergence of two rivers. 'That is where I would like to have this interview. Watching my father's' hands knitting his fishing gear. The smell of turmeric, chilli, and coconut. The simplicity of my mum and dad discussing spices and him leaving to the river with the unspoken

Fig. 19.2 Fisherman inspired by Maimunah Mohd Sharif's memories and preferred location for this interview

promise to be back in less than an hour with fresh fish. The understanding of living in harmony with nature, without needing the words for it.' (Fig. 19.2)

The convergence of two rivers creates particular significance in traditional Malay culture, representing the meeting of different forces that together create something with more power than either could achieve alone. In Islamic cosmology, rivers represent divine blessing and the source of all life, whilst in local tradition, the meeting of waters creates a sacred space where different communities can come together whilst maintaining their distinct identities. For Maimunah, this convergence embodies everything she believes about sustainable urban development: cities function best when they bring together diverse knowledge systems, when rural and urban wisdom inform each other, and when traditional and contemporary approaches create synthesis rather than replacement.

Here, her mother would show visitors how to grate coconut for cooking and how to use every part of the palm tree for different household needs, whilst her father, a religious teacher who served as her greatest role model, could demonstrate the fishing techniques that provided protein for the family without depleting the river's abundance. These are simple yet sophisticated approaches to sustainable living that integrate spiritual values with practical knowledge, in the understanding that human well-being depends on maintaining reciprocal relationships with natural systems. The fishing practices her father taught embodied principles that would later inform her approach to urban development: taking only what is needed, ensuring that extraction does no't compromise future abundance, understanding seasonal cycles and respecting them, and recognising that individual actions affect collective resources.

'We talked about water from the well, we took water from the river. We don't know what is pollution at that time, because the water is crystal clear,' she reflects, her voice becoming emotional as she describes a childhood where sustainability was no't a policy framework but a way of life. The rivers of her childhood ran clear because upstream communities understood their responsibility to downstream users, because traditional agricultural practices worked with, rather than against, natural cycles, because waste was minimal and entirely biodegradable, because population density remained within ecological limits, and because spiritual values reinforced practical conservation. 'We don't know what is waste. We don't know how to throw rubbish into the river. We already recycled everything because everything in the village has a use.' This early education in what would later be called circular economy principles happened without grand theories. Nothing was wasted because waste represented both economic loss and spiritual disrespect for the gifts provided by natural systems.

The abundance of her childhood was measured not in accumulated possessions but in the reliability of natural systems and the depth of community relationships. The simplicity did no't create poverty but rather abundance: of time, relationship, and connection to place that contemporary urban life often struggles to provide, despite much greater material wealth. Yet all of it seemed like a disadvantage when Maimunah received a scholarship to study town planning at Cardiff University in Wales, creating what she describes as a profound cultural shock, where the transition from village life to British academia meant 'I don't know what is broccoli. I don't know what is cauliflower. I don't know what is potato because the food that I eat is totally different.' More challenging than unfamiliar vegetables was learning to navigate technologies and social systems: 'I don't even know how to turn on the gas stove. It's something very, very new to me.'

The cultural displacement extended beyond practical unfamiliarity to deeper questions about progress, development, and what constitutes a good life. British academic culture assumed that rural Malaysian village life represented limitations to be overcome rather than wisdom to be preserved and adapted. The sophisticated knowledge systems her parents had taught her were not recognised as relevant to town planning education.

Among her scholarship cohort, all from urban Malaysia, she felt isolated and insecure. 'I was the only one from the village. "Why I am born in this family?" I wondered. But now looking back, I thank my parents that I had the opportunity to grow up the way I did, be able to go and study in the UK, be exposed to this environment, and come back to my country with both new and old eyes.' This initial sense of inadequacy would prove to be a profound misperception, as her rural background provided exactly the kind of integrated understanding of sustainable systems that contemporary urban development desperately needs.

She recounts with self-deprecating humour what she calls 'Maimunah's mistake'—a story about development, progress, and the seductive power of modernisation. Returning to her village with her planning degree, she felt compelled to demonstrate that her education had transformed her and that she now understood what progress looked like. Judged by Western standards, her family's traditional house built by her great-grandfather appeared primitive. The traditional house, however, incorporated natural ventilation that eliminated the need for air conditioning (AC), materials sourced from local ecosystems that created comfortable indoor temperatures, integration with the landscape that provided both beauty and function, and spatial arrangements that supported extended family relationships whilst maintaining privacy—solutions developed over generations creating architecture that worked with local climate, social structures, and ecological systems.

'When I came back, I said, "I don't want to stay in this house anymore. Enough is enough. I want to change because I came back from the UK. I want to show what progress looks like to the community. I want to build a brick house."' The house she built incorporated everything she associated with advancement: multiple rooms that prioritised individual privacy over family connection, indoor plumbing that eliminated traditional composting systems, concrete and steel construction that required imported materials and fossil fuel energy, and crucially, air conditioning. Broadly smiling she asks us: 'Do you think I'm crazy? The first house with AC to show that Maimunah came back from the UK and graduated in town planning. The whole village came to see my house. They wanted to be in my rooms to feel the air conditioning for so many weeks.' The demonstration effect proved more powerful than she had anticipated. The air conditioning represented more than comfort; it symbolised connection to global modernity, proof that rural Malaysia could participate in the same lifestyle that dominated cities worldwide.

'Now if you go back to the village, so many people in the village are building brick houses. It's my mistake that brought a very bad example. They should be keeping the traditional home.' The unintended consequences of her personal choices—multiplied across the community—demonstrated how individual actions ripple outward to affect entire systems and how what appears as personal progress can become collective regression when measured against ecological and social sustainability.

The irony was not lost on her: The very education that was supposed to prepare her to improve villages had led her to undermine village wisdom, demonstrating how formal education often creates experts who understand less about sustainable systems than the communities they are supposed to serve. This recognition led to her current plan to rebuild her family's traditional house using contemporary techniques that can maintain traditional design principles whilst incorporating modern amenities. 'I want to rebuild my father's house. I know the carpenters who can still do that. I want to demolish the brick house. Go back to what today would call a nature-based solution.'

'Later, when I was mayor in Penang and subsequently at UN-Habitat, I worked with the schools on waste management and on preserving traditional knowledge. Realising that 40% of the budget went to managing waste, I launched the Waste Wise Cities Initiative [in 2018] and started promoting "the four Rs": rethink, reduce, reuse, recycle. Whatever you do, you think before you buy things, before you go to the restaurant, before putting things on your plate. Eat till finish, so that you don't waste.' This approach applies village wisdom to urban contexts, recognising that sustainable cities require cultural transformation as much as technical innovation.

Her six years leading UN-Habitat provided countless opportunities to witness how these principles could transform urban life, with stories documented in her 'Building Bridges' report demonstrating how small interventions rooted in community needs create profound change. 'I remember this formal visit to Afghanistan, just months before the Taliban takeover. For security reasons field visits could only take a handful of minutes, but I witnessed things that justified all the complexity of the work. I remember meeting this young girl, expressing her gratitude for something as simple as a staircase.' Prevented by security from approaching directly, the girl managed to convey through translators her appreciation for infrastructure that connected her hilltop village to the road below. 'What we did was just to have built a staircase. She kept saying thank you, and for giving her "a little brother." I said, "What do you mean by 'little brother'?" She said: "When my mother was pregnant, she normally could not go to hospital because we cannot bring her down to the road. When my mum needed to deliver my little brother, we managed with the staircase to carry her down and take her to the hospital."' The impact of a staircase costing less than $20,000 on the family and wider community proved immeasurable, illustrating how development work succeeds through attention to basic human needs.

In Iraq, at Yamouk Park, she presided over the opening of public space created on land cleared of landmines, where 'I was very, very unhappy and sad to see children with one leg, with no leg at all, one hand, trying to run towards me and hug me to say, "Thank you for giving us this park. And now we can use this park without any problem thinking about landmines."' The project represented more than environmental clean-up: it embodied the possibility of healing from conflict and of transforming spaces of violence into spaces of community.

In Kenya, visiting the Mathare slum, she encountered a 13-year-old girl whose request for a present surprised her: 'I thought she would ask for shoes or a bag. You know what she asked? She asked me, "Can you supply me with sanitary towels? Every month, I will miss at least two weeks of school because I cannot go to school when I have my periods."'

The encounter led to immediate action—Maimunah calling her husband to purchase supplies—but also to a long-term relationship: 'That was in 2018. I'm still keeping in touch with the girl, and now she's in training college, and I hope that she can be a teacher one day.'

Talking to Maimunah, one realises she could go on for hours recounting these all-too-human anecdotes and leave you with a totally different reading of sustainable development—far from conceptual but humanising values like dignity and the power of just listening before acting, understanding local contexts before proposing interventions, and recognising that the most sophisticated solutions often appear deceptively simple.

As she prepares to leave UN-Habitat, Maimunah takes pride in having transformed the organisation from an $8 million deficit when she arrived to a $6 million surplus at her departure, growing the agency's portfolio from $150 million to $1.1 billion, but the numbers matter less than what they have enabled: thousands of small interventions that result in what she calls 'development with a human face.' When asked which teams really make the difference, she replies, 'I would like to have people who are humble, have humility, and are yet visionary. And have positive thinking.' She says that when presented with new ideas, even seemingly impossible ones, she wants colleagues who respond with 'Let's think about it' rather than immediately saying 'It cannot be done.'

She returns to where she started the conversation, to family: her parents, who provided a foundation in sustainable living and spiritual values that enabled her to navigate complex institutional environments whilst maintaining commitment to community service. To her husband, who always said, 'You are the best, be yourself. There must be a reason why God chose you to be there,' who supported her decision to take the UN position despite the personal costs. To her daughters, who encouraged her to accept the role as 'mayor of the world' rather than retiring as mayor of Penang, and to her new grandson who represents the future she works to create.

This spiritual dimension of her work reflects Islamic values about service, responsibility, and the interconnection of all creation that informed her childhood education by the rivers, understanding that individual talents and opportunities create obligations to serve collective well-being rather than purely personal advancement. The theological framework that guides her work emphasises stewardship rather than ownership, service rather than domination, and humility rather than pride.

20

Mark Watts (UK): Finding North in a Climate Crisis

Abstract Mark Watts is Executive Director of C40 Cities, a global network of nearly 100 mayors of the world's leading cities united in confronting the climate crisis. C40 Cities represents 25% of global GDP and hundreds of millions of residents. Named a 2023 TIME100 climate leader, Mark helped establish C40 in 2005 and became Executive Director in 2013. Previously, Mark was a director at Arup and climate advisor to the Mayor of London.

Keywords Urban climate action • City networks • Transformative leadership

30 January 2024, Online from London, England; Updated December 2025

It is a winter morning in the UK when Mark Watts, clearly comfortable with connecting across continents and time zones, joins us from his London home. There's an unexpected intimacy to finding him in this domestic setting rather than his usual office at C40 Cities, where he leads

C. Oke, J. Verbeeck, *City Leaders for Sustainable Futures*,
https://doi.org/10.1007/978-981-95-9544-0_20

the world's most influential network of climate-focused mayors. He has spent the better part of two decades connecting cities across the world in their fight against climate breakdown but is not someone to be tempted to speak about the work done; he would rather talk about how many challenges the journey still holds. He speaks with the same passion about carbon reduction targets as he does about cycling superhighways, demonstrating how the work of urban transformation is simultaneously technical and profoundly personal (Fig. 20.1).

He apologises for the noise of construction workers in the background, and if teleporting were an option, he would take us for a walk to a place nearby: a windswept hilltop in Greenwich, London, where one finds the Royal Observatory, the starting point of the Prime Meridian. It is one of the few places in London with natural elevation, where the sprawling metropolis reveals itself in all its complexity and possibility. Here he finds both reflection and inspiration. 'Of course there's the privilege of all the places I've been to, both great and small, but actually I'm happy to keep

Fig. 20.1 Mark Watts, image inspired by photograph by C40 Cities Leadership Group

London as the imaginary setting for our talk,' he says. It is still the place he calls home, where intimate knowledge of place offers an anchor point for an understanding of global urban challenges.

'It's looking sort of south to north through the east of the city. It's a part of the city that has hugely changed. I was lucky to be part of the mayoral administration that really kicked that off through the development of the Olympics,' he explains. From the hilltop, the abstract concepts of urban planning become visceral and visible. But Greenwich Hill offers more than just a panoramic view; it carries deep historical resonance. 'Three hundred years ago, London's air was clear enough to do serious astronomy, right in the centre of London,' Mark observes, with a mix of wonder and subtle reproach.

'I'm a historian by background, but astronomy was one of my first loves when I was a teenager, spending my evenings in the back garden with the binoculars or telescope. Now imagine, when you walk into the Royal Observatory, the first thing you see and can touch is a 4-billion-year-old piece of meteorite that fell to earth. That sense of time, the age of the earth, and that very, very small part that humans have played in the history of the planet that we live on.'

This perspective—simultaneously humbling and galvanising—runs through all his work on climate change, providing both context for the crisis and urgency for action. The hill also serves as a place of personal ritual and renewal: Mark associates it with Saturday mornings when he's trying to 'shrug off all of the stress of the work week' before visiting one of his favourite museums at the base of the hill (the National Maritime Museum). This is the same hill where he has taken his nieces and godchildren over the years, watching them discover the simple joy of rolling down the big slope and taking in views of the whole city, creating personal memories that layer meaning onto the landscape.

Mark's journey to climate leadership began not with grand ambitions but with a book and a mayor's casual instruction. Growing up in Burton upon Trent—a town whose very name means 'beer town'—he brought a historian's perspective and a socialist's commitment to collective action to his work with Ken Livingstone in the early 2000s.

Then-mayor Livingstone would occasionally give him books to read and summarise. 'One of the books he gave me—if I had it, this would be my object—is Jared Diamond's *Collapse*,' he shares. 'After digesting, Ken asked me, "Were you scared?". I said, "I'm terrified. What is this climate change thing? This sounds awful." "Well, that's good," he replied, "because you're now my climate change adviser."' The casual nature of this assignment—'I know nothing about it,' Mark protested, to which Livingstone replied, 'Well, neither does anybody else. So go out and learn'—speaks to both the emerging urgency and the improvisational nature of early climate leadership.

The knowledge existed, but the institutional structures had not yet formed. Entire policy directions are sometimes shaped by the tiniest individual initiative. In this case a mayor decided that climate change needed attention and assigned someone to figure out what that meant.

The object that for Mark most represents this transformational period is an old-fashioned compass, found in a junk shop in Walthamstow and carried with him on journeys to cities across the world as he sought to understand what climate action looked like in practice. This was the pre-smartphone era, when navigating unfamiliar cities required actual maps and the ability to determine direction. 'The hardest thing is just working out which way, which direction you're facing,' he explains. 'And if I could work out north, east, south, west, then I could work out from the map where I needed to go.' It is at the same time a metaphor for his approach to climate leadership—the understanding that before you can chart a course towards solutions, you need to know your position, what direction you are facing, and how to orient yourself in unfamiliar territory (Fig. 20.2).

Reading *Collapse*[1] fundamentally shifted Mark's worldview, challenging the materialistic socialism of his core politics with a more ecological understanding. 'I think my perspective was one of concrete conquering nature,' he reflects. 'And I think reading Jared Diamond was for me the start of what's been a long journey of understanding humanity as just another part of nature, part of an ecosystem.'[2] The book's most disturbing insight was the possibility of collective self-destruction. 'Human civilisations were and are able to just get into a collective malaise where they

[1] Diamond, J. (2004). *Collapse: How Societies Choose to Fail or Succeed*. Viking Adult.

[2] Collapse by Jared Diamond.

Fig. 20.2 Illustration of an old fashioned compass, described by Mark Watts

destroyed the very ecosystem that made life possible,' Mark explains, 'with nobody at any point intending to do so but everybody collectively ignoring all the signs.'

This understanding drove him to seek out cities that were already taking climate action seriously, leading to journeys across the globe with his compass in hand, learning from places like Stockholm, Copenhagen, Toronto, San Francisco, and New York. What he found was a scattered network of urban leaders who had been thinking about these challenges for years, developing approaches that combined technical innovation with social vision in ways that national governments seemed unable or unwilling to attempt. 'What I immediately thought was, "Oh my God, here are people who've been thinking about this for a decade or more, way, way ahead of us,"' he recalls. In Scandinavian cities, he encountered a different model of society—one that had emerged from the oil price shocks of the 1970s with a commitment to efficiency and long-term thinking that produced higher building standards and integrated transport systems.

Mark recalls Nicky Gavron, then-deputy mayor of London, asking him, 'Why don't we just keep doing this?', as Mark compiled London's first comprehensive climate action plan. 'Why don't we keep talking to each other? Because this is going to be a tough and a long journey. And let's create a network of cities that's just focused on the climate crisis and that can help each other tackle it.'

Mark speaks about Ken Livingstone's leadership as the kind that makes urban transformation possible: visionary yet practical. 'He was always looking 8 to 10 years into the future about the direction that he wanted the city to take,' Mark reflects, 'and yet at the same time, the most brilliant administrator. You knew if you went into a Cabinet meeting with Ken that he would have read the papers more meticulously than anybody else in the room.' This combination of long-term vision and operational excellence created an environment where ambitious projects could actually be implemented and where ideas like the congestion charge could move from concept to reality despite significant political resistance.

Livingstone's management style was equally important: 'Here's what I want you to do. Go away and do it. No need to come back unless you've got a problem. If you've got a problem, I've got your back and, I'll help you sort it out.'

Now, two decades later, Mark finds himself in the position of fostering similar leadership across a global network of cities. His optimism, he explains, comes fundamentally from collaboration: the daily experience of working with people who refuse to accept that the climate crisis is insurmountable and who continue to develop innovative approaches and learn from each other's successes and failures. 'I feel constantly optimistic and inspired, but I think that's fundamentally because I've been in the privileged position of collaborating with lots of people, and collaboration breeds optimism because you see you're not alone.'

This collaborative optimism is, however, tested constantly by the scale of the challenge and the persistence of opposition. Mark is clear eyed about the reality that despite decades of progress on sustainability technologies and policies, 'carbon emissions are still rising. We're still pushing past multiple planetary boundaries. We're still destroying ecosystems left, right, and centre.' The gap between knowledge and action, between technical solutions and political will, remains frustratingly wide.

Yet he finds energy in what he calls 'the struggle': the recognition that transformation requires not just inspiration and collaboration but also the willingness to 'overcome and defeat those people that are trying to take us backwards or hold us back.' It is the stubborn optimism Christiana Figueres talks about: a commitment to continue working for change even when progress seems impossibly slow.

Just like stubborn optimism, music can be a cure for the pain. The conversation shifts to the Glastonbury Festival, a temporary city of nearly 300,000 people that emerges each June in the Somerset countryside. 'Everybody changes when they're there,' he observes. 'Just the love and care that everybody has for each other.' Glastonbury represents everything Mark believes cities can and should be: diverse, creative, and collaborative, where strangers become community and where music provides the soundtrack for collective transformation.

This vision of cities as spaces of encounter and possibility runs through all of Mark's work. He sees cities as humanity's greatest collaborative achievement. 'The ability to collaborate made the difference,' he reflects. 'We weren't the strongest species, we've never been the fastest. Maybe the cleverest. But it was that ability to collaborate that really differentiated us.'

The climate crisis becomes not only an existential threat but also an opportunity for collaboration. 'We now have the knowledge for the first time in human history that we can see the future. We can understand what is going to happen with great certainty if we carry on living the way that we are.' This knowledge creates both responsibility and possibility.

Mark's vision of urban transformation combines top-down leadership with bottom-up community engagement, technical innovation with cultural change. He points to examples from across the C40 network: air quality improvements in Chinese cities driven by popular anger about pollution, flood-resistant urban farming projects in Manila, and the cycling infrastructure in London that has transformed both mobility and urban culture.

'You need a combination of at least two things,' he says. 'You do need visionary, determined political leadership. All the amazing big changes I've seen in my lifetime around cities couldn't have happened without that visionary, strong leadership.' But visionary leadership comes from acute understanding of society, and political leadership alone is not sufficient: 'It's equally got to translate through to thousands of small communities working together and individuals thinking about what they're doing in their family or on their own to change their own lives.'

This understanding of change as simultaneously structural and personal runs through everything Mark has learnt about urban transformation. Cities change through countless individual decisions and actions, but those are shaped by infrastructure, policy, and culture.

As our conversation draws towards a close, Mark reflects on the kind of leadership team he would want for rebuilding society in case there would be a climate collapse: a fantasy rock band of urban transformation that includes political leaders like António Guterres and mayors Sadiq Khan, Yvonne Aki-Sawyerr, and Anne Hidalgo, economists like Mariana Mazzucato and Kate Raworth, and climate leaders like Christiana Figueres. But he acknowledges with characteristic humour that his team lacks practical skills: 'There's nobody in that list that's going to build a shelter. There's nobody who's going to help us work on a farm. We're going to struggle to eat, but we're going to have lots of good conversations.'

21

Angharad Wynne-Jones (Australia/ Wales): Where the Land Meets the Ocean

Abstract Angharad Wynne-Jones is a Cymry (Welsh) Australian living on unceded Wurundjeri Woi Wurrung lands in Naarm (Melbourne). As Facilitator of Creative Climate, a culture and arts climate action peak body, she delivers sector support and systemic transformation. She was Artistic Director at Arts House Melbourne and founded Tipping Point Australia to energise cultural responses to climate change.

Keywords Climate justice • Cultural transformation • Social-ecological systems

7 February 2024, Online from Brunswick, Melbourne, Australia; Updated December 2025

Angharad Wynne-Jones has spent years in 'edge zones,' those places where the constructed city meets something wilder and more enduring, where the urgency of the climate crisis meets the transformative power of arts and culture. Having lived in several places in the UK and Australia, she acknowledges that we are currently speaking from the lands of the

C. Oke, J. Verbeeck, *City Leaders for Sustainable Futures*,
https://doi.org/10.1007/978-981-95-9544-0_21

Fig. 21.1 Angharad Wynne-Jones

Wurundjeri people of the Eastern Kulin Nation and honours elders past, present, and future—a practice that reflects her understanding of how settler culture must reckon with both its colonial legacy and its responsibility for creating more just relationships with Country (Fig. 21.1).

It's February 7, summertime in Melbourne, when we virtually meet Angharad in her Brunswick home office. The sounds of inner Melbourne drift through: trams and conversations, the energy of a city that never quite stops moving. Angharad is part of the reworlding team, 'a collective of Indigenous, people of colour, settler and LGBTIQA2S+ artists, scientists, thinkers, and change-makers with a track record of collaboratively working at the intersections of art, the climate emergency leadership, speculative futures, and disaster resilience.'

Angharad is currently co-facilitator of Creative Climate, new national peak body for arts and climate funded by Creative Australia. At the time of our interview she was working at the State Library and having a new-found respect for the book as an object of knowledge transmission alongside aural storytelling traditions. When she jumps into the conversation, she stages a wonderful play of the many crossroads she has encountered between the intimate work of cultural change and the vast challenge of planetary transformation.

'Half Moon Bay.' She does not have to think long when asked to choose the preferred location for this conversation. Not too far from Melbourne, it is a stretch of coastline where Bunurong and Boonwurrung lands meet the water. 'It's on the edge of the city where the land meets the ocean,' she explains. 'Yet despite being deeply connected to the city and somewhat hidden, it's pretty much the first beach in the bay where you have the sense of the rising landscape of the red cliffs. There's something really magical about the place, as it both kind of shelters and supports but also invites to see the big horizon. That's maybe something in a city that is critical.' Her visualisation is as much about the place itself as about those intersectional spaces between the natural and the urban, the immediate and the eternal, the human and more-than-human worlds.

The horizon proves not to be accidental. Over the years, Angharad has seen the most profound transformations happen not in the centres of power but at the edges, where different systems encounter each other. She is, however, not the person for large theories. Angharad draws us into the story of the weedy sea dragon, for which the bay is its natural habitat. As the males carry the eggs, she describes the creature with evident delight as having 'been doing gender diversity way before the rest of us.' She observes how nature offers models for ways of being that challenge conventional assumptions and that in turn flavour her art and activism.

'I have an abiding love of swimming and being in water,' she says. Interviewing Angharad is not so much a question-answer pattern. Rather, she brings a certain flow when she speaks, allowing her words to float, to carry, to crash, and to whirl around. Ever since growing up in England and spending time in Wales, where you are never far from lakes, rivers, streams, or the sea, water has always been part of her life. 'Since childhood, I find being close to bodies of water, as I'm sure we all do, incredibly rejuvenating. They are critical to mental and physical health.' Later in Australia, when moving from Gadigal land in Sydney to St Kilda in

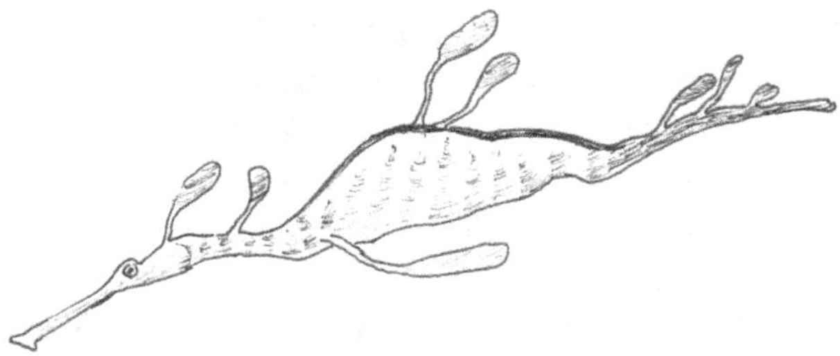

Fig. 21.2 A weedy seadragon, representing many lines of inspiration for Angharad Wynne-Jones

Melbourne in 1997, water again had to be part of the picture. Half Moon Bay became an intentional pilgrimage, a destination for special occasions rather than daily routine (Fig. 21.2).

Apart from the connection to the sea, the Bay crystallises Melbourne's geography and history. 'The colonial project was very busy stamping out evidence of people that have lived here for millennia, and busy blowing up river beds,' describing how the Birrarung (Yarra River) was originally 'a series of ponds' before colonisers 'blew up the rock formations between the ponds in order to bring shipping closer to the ports and in the process destroyed the function of that river.' After a pause, she says, 'There's a lot to recover and remember.'

She often returns to encounters with Indigenous perspectives and practices. 'Every public event that happens in Australia—particularly in the last five or 10 years—a Welcome to Country is expected.' These represents something profound: 'It doesn't matter what the event is, there will be an elder who is stating that sovereignty was not ceded, that the land represents an unbroken culture for tens of thousands of years. These Welcomes puncture whatever fabric of the system is operating, serving as a great leveller where the elder is essentially saying: "You think you've got something really big going on. I'm just telling you, we're here and this is our land, and you need to have a relationship to it."'

This leads us to an object she carries that represents her sustainability journey. She shows a piece of pounamu (New Zealand greenstone) given to her by Lemi Ponifasio, a Samoan-Māori artist whose company Pau she encountered during her time as Artistic Director of the London

International Festival of Theatre. 'I wear the stone whenever I feel in need of courage and authority,' she says. 'Traditionally you couldn't take this for yourself. You had to receive it. The greenstone represents a really strong sense of a different way of transacting: one that is about acknowledging the relationship and responsibility to each other.' The gift came in the context of Lemi's reworking of *The Tempest* retold through Samoan and Māori perspectives. For Lemi's culture, 'every rock, every—what Westerners might call an inanimate—object, has a spirit and is animate, and is deeply connected to human beings, to each other and to other species.' It helped Angharad question her worldview and opened her to ways of thinking about relationship and responsibility that extended far beyond human-to-human connection.

Her time in London brought other encounters. One was with an organisation called Tipping Point, set up by Peter Gingold to bring scientists and artists together. 'At that point, the conversation in Australia had already become politicised. It had stayed there for a decade or more and just faltered. But in the UK it wasn't yet politicised in the same way.' Where Australia's climate conversation had become toxic and polarised, there was in the UK 'an appetite for knowledge exchange, where scientists who had been sitting on information for decades wanted to share what they knew and sought out communicators, makers, creators to help translate their findings.' The initiative proved a personal tipping point. She had not seen 'how it could spill into professional working life.' The experience gave her permission to recognise that 'it's a ludicrous thought that we can divide all these things in life and completely antithetical to what art and culture are about. No change happens in isolation. It has to happen collectively, it has to happen with conversation, with artistic support and imagination.'

Another crucial influence was Matt Wicking, whose day job was in environmental sustainability but who after hours was also a skilled singer/songwriter. He showed Angharad new ways of how personal and professional transformation could integrate but also what she calls 'the dark depression of working in sustainability.' Matt connected her to the Dark Mountain project in Wales, which helped her understand the importance of grieving as part of ongoing engagement. 'How do we grieve? How do we let ourselves grieve so that we can continue?' she asks.

Moving back from the UK to Australia, she kept building on different generational approaches. She points to younger-generation activists like Lana Nguyen and Eliki Reade from A Climate for Art and how they are 'taking the campaigning methodology and making it not only intersectional but also relational.' This new generation, she observes, has 'grown up with climate change as knowledge in ways that fundamentally shape their approach.' She clarifies, 'They're not burdened in the same way as boomers by the notion of sacrifice, even though they have it far tougher than my generation as far as cost of living, secure housing, and, of course, the climate crisis.' Instead, these younger activists are finding ways to live more simply, move away from consumption, and navigate complexity.

Over time, Angharad's vision of transformation became fundamentally collective rather than based on individual leadership. 'I don't think the change is located in one person or one action,' she reflects. 'It's absolutely about the connections between people, our world, and our planet.' The challenge is 'working in small contexts, in locally specific contexts whilst finding ways to share knowledge that can be useful elsewhere.'

She returns to grieving when confronted with sustainability paradoxes, not least in the context of her international work requiring frequent flying. She shies away from easy answers. 'We need to hold those kinds of complexities and paradoxes in our own actions,' she says. 'We need to find a better way of failing.' She's not looking for ideological purity but seeks ways to 'sit or walk with that complexity whilst continuing to act and create.'

We wonder: What would a festival at the brink of civilisation collapse look like? There are both simple joy and depth in her reply. 'Obviously it would be a lot of queers having a fabulous dance party,' she playfully shares and then muses: 'I would invite all the people that are currently planting seeds as part of tree projects all over Victoria to bring their beautiful seedlings and plant them all over the city. Not so much spectacle, more shared space on the same level, doing things together, with a lot of great dance music and costumes. And of course, led by our First Nations elders who have been doing it for millennia for exactly that reason.'

Provoked by our wondering whether cities themselves are a kind of festival, she reveals the complexity of her relationship to urban life: 'I think the city is a festival in preparation. We still need and seek moments of coming together, sometimes just witnessing each other and being around that kind of collective energy.' The city becomes a place for both preparation and arrival, a work in progress, laboratories for new forms of relationship: whether we can learn to live together in ways that sustain rather than destroy the conditions for all life to flourish.

22

Conclusion: The Edges We Walk and Song We Share

Abstract Just as our initial conversations on inspiration and interfaces led to this book, our current conversations venture further into the boundary spanning landscape. We did not set out to work on the edges, nor are we the first ones to dive into this concept. We did not always have the words or a name for what we were doing. We only knew that without it, things fell apart faster. The systemic challenges cities carry will not be answered by individual boundary spanners doing good work in parallel. They require us to genuinely build together. Walk with us. Reach out to us.

Keywords Boundary spanning • Collaborative governance • Transformative leadership • Urban resilience • Capacity building

Dubai, 2024, Updated January 2026

Just as our initial conversations on interfaces led to this book, our current dialogues venture further into the boundary spanning landscape. We did not set out to work on the edges, nor are we the first ones to dive into this concept. Most of us were trained for something else: research, analysis, planning, engineering, policy, and social work. We learnt the languages

C. Oke, J. Verbeeck, *City Leaders for Sustainable Futures*,
https://doi.org/10.1007/978-981-95-9544-0_22

of our fields and tried, earnestly, to apply them. What we were not taught was how life, community, and everything urban all live in the gaps between those domains and their distinct languages and traditions. At first, it felt like confusion. You would leave a meeting knowing something important had been said but also knowing that you had barely scratched the surface or that it would not travel. Or you would carry a concern from one talk to another, only to realise that it had lost its meaning in translation. We did not always have the words or a name for what we were doing. We only knew that without it, things fell apart faster.

Over the past years, like Leopold Bloom wandering Dublin on 16 June 1904, we have followed contemporary urban wanderers in their odysseys. James Joyce wrote that by getting to the heart of Dublin, he could reach the heart of all cities.[1] Though there are no two cities the same, they also have something universal. Joyce rewrote Homer's story from the perspective of an 'ordinary' hero. The meaning in his narrative emerges through attention, not triumph. With Leopold Bloom, he turns the myth inward and demonstrates the complexity of urban life as the essence, not the obstacle—all unfolding in a single day.

As with the Odyssey, only as we wrote did our book find its own proper voice. The interviews all took their own direction, as did the meandering questions between the two of us: Who are we writing for and to what extent? Can we sufficiently evoke the quiet and invisible work that has stuck with us and that we have sometimes also contributed to? Some stories are hard to tell. Can we open up the conversation about heroism—that it is not about fixing everything but about coordination? That responsibility lies in discovery and care, not domination. Popular culture today is already moving from heroes to teams. Can we learn to tell better stories—ones that train people to collectively hold responsibility and institutionalise sense making and boundary spanning? To what extent do we ourselves filter what we see? What are our own roots and experiences that sometimes drive a dialogue to a certain direction? Can we stay oriented without certainty?

[1] Joyce, J. (2010). *Ulysses*. Wordsworth Editions.

Roots

One set of footprints starts on the Sydney Harbour wharves, where a visiting six-year-old, animal-loving girl watched ferries cut through turquoise water. This awe of the natural world lapping at her feet instilled a strong sense of care, alongside a love of exploring new places with family in tow. Venturing to the top of Australia to study the marine world that she had discovered at an early age was exciting, as was London, where her postgraduate studies in fisheries conservation led to skills in communicating complex challenges by engaging and listening to people from all perspectives. It also led to an understanding of decision-making and a seat as an elected city councillor back in her hometown of Melbourne. The other begins in Antwerp, where a young, rather introverted teenager first discovered urban stories in the library where he waited for his parents after school. Classics of literature and graphic novels filled his afternoons. Classic stories, science and social fiction, and mythology helped him to rethink his surroundings: a social housing context and a risk-averse education that baked in boundaries by design. Growing up in a wealthy Western country where rule of law had quietly substituted for much of the moral compass, the instinct to cross boundaries had to be found—it was never assumed. Later on, his first job proved to be his real training: as a young political adviser to the vice mayor, he was asked to shadow the full breadth of city policy and find the connections between domains—to help make sense of the city as a whole. Two formative stories separated by hemispheres, connected by the same question: How does this world work, and what is our place in it?

We met as colleagues, connecting on a similar experience of having both worked in urban politics, urban development, and global city networks. Science and research, public-private multi-stakeholder action, and arts and popular culture have enriched both our trajectories. We became friends, anchoring on the question of how to connect intentions to actions and how to avoid only talking about cities rather than seeding the rich knowledge to help strengthen local communities that do the actual implementation and steward trust and resilience. We did not want our collaboration and wanderings to end when formal employment did. This book therefore is the continuation of our conversations but also the invitation to help in rethinking how we bring radical change agents together and organise long term co-creation that goes beyond individual agendas and perhaps even organisational boundaries to hold complexity and multitudes.

During the 2021 Innovate4Cities conference, held fully virtually during the pandemic, we facilitated 24 hours of conversations across time zones: an urban Decameron. One question kept surfacing: Why do ever more compelling scientific insights not lead to more action? We are believers in storytelling and oral traditions. And in some of the mythology we have lost. We consider them not messages to persuade but forms of boundary spanning that helps people to think and act together. Each of our conversations with our 20 storytellers could be produced as podcasts or longer stand-alone pieces, but we chose to find added value in bringing these stories together as a collective and build from there.

The Objects We Carry

When we asked each other what object would remind us of the spark that started our sustainability journeys, we discovered we do not really have objects. We have visual ways of thinking, sensory memories that transport us back.

For Jorn, it is the long early-morning walks when visiting a new place, watching the city wake up and feeling its pulse, spotting the vital signs of the city as urban metabolism, sensing the continuity across time and the respect for institutional memory as shared in the local bakery, coffee, and grocery shops. It is trying to watch with new eyes and listen without filters. He remembers a trip to South Africa and how he experienced the visceral weight of the past, present, and future. For a Belgian, Afrikaans uses a lot of metaphors and offers a very visual understanding of reality but also carries a lot of trauma and transition. The land struggles but also holds the dark and bright sides of life and humanity. He remembers one night in Cape Town, walking in the dark with Western confidence, being suddenly surrounded, experiencing deep fear, running fast, and having the bed-and-breakfast owner eventually saying: 'Take off your Western view and see the country as it is.' That moment taught what culture, community, and history mean when trying to understand urban life. You can take nice pictures, but do you really (care to) understand? How to rebalance the institutional, cultural, and creative parts of our society and allow a literacy of civic complexity?

For Cathy, it is a photo capturing her many memories of falling in love with the ocean and nature and finding lifelong joy exploring new cities with family. It is standing with her brother, sister, and parents in front of spectacular Sydney Harbour and feeling a connection to the natural world that came before any formal understanding of sustainability. Every time she sees that harbour, the moment returns: the smell of ferries and fish and the wharves that were playgrounds. Her path through marine science led to communications work at an NGO in London, and to St. James's Palace for a fundraising event where Mark Kurlansky—who wrote *Cod*—was the guest speaker. There was a game, designed to raise awareness of the connection between global trade decisions and fishing stocks. As captain of the losing team, Cathy collected the prize: a plush toy fish, presented by Prince Charles himself, now King Charles III. That toy fish still sits on a shelf in her study, a reminder that she once lost all the fish in the world in a game and that stories about overfishing matter—not just metrics and data but hearts and minds and livelihoods (Fig. 22.1).

Fig. 22.1 An old Sydney Harbour ferry, representing Cathy Oke's sustainability beginnings

Staying Close to Home

Both of us, like many people we have invited for this book, are global citizens who keep a close connection with the land and environment they call home. This sense of community and place—the grounding from knowing where you started and where you can always return—runs through these stories like an underground river.

Growing up in a family of mostly civil servants, Jorn saw the small rituals and gestures, hopes, and fears in a warm environment of moral duty and care. His grandmother taught him sustainability by intergenerational grandparenting, by connecting him to global wildlife fund initiatives and publications, stories of the wider world. She grew a community growing older herself. She was able to make something out of nothing. She did not need a circular agenda: she already gave a third and a fourth life to everything. Sharing the material and immaterial probably lengthened her life. Jorn was the first in his family to attend university. He ventured out with a mix of dedication and unsettledness, finding his own voice and tribe while wrestling with a question that would take years to settle: what 'Indigenous' and belonging means when you are a Western white male who has outgrown his neighbourhood but not yet found a new one. These wanderings brought new friends and new worlds, but also the particular kind of loss that comes when you grow in directions old friends cannot follow. Yet, the more you stray, the more you need to anchor someplace. For a number of years now, he has been helping build a cohousing project just outside of Brussels into 'a third place': a living and open community and new form of enlarged family (Fig. 22.2).

For Cathy, family has always been central to that sense of home. It is no coincidence that she, her sister, and her brother all live within a short distance from their parents, who still live in the family home in inner-city Melbourne. That gravitational pull towards family and place—the ability to have frequent meals together, to be close by for the daily small moments, not just the big ones—it is a choice about what matters when you strip away the professional achievements and global travel and come back to the people who knew you before you became whoever the world thinks you are.

Fig. 22.2 An illustration of the Nello and Patrasche statue Antwerp, Belgium, near Jorn Verbeeck's home

This is the spectrum these stories offer. Look at where I am now. But also look at where I began.

Rhythm and Belonging

There is another thread running through our conversations and through these stories that we have not fully explored yet—one that deserves its own book: music. The rhythms and lyrics of cities, the soundtracks that shape urban experience, and the musicians who translate the chaos and beauty of place into melody and verse.

For Jorn, music has always been inseparable from understanding the world. The literature he studied taught him to steal thoughts from across time, but music taught him to feel them. It is there in the record stores he mentions wandering through, in the albums that cannot be listened to anymore in their entirety by kids frantically scrolling streaming services.

Music carries memory the way smell does: a particular song can transport you back to a sometimes long-forgotten place with visceral completeness.

For Cathy, many of her favourite moments in cities happen when live music fills not only venues but also beautiful outdoor spaces. These include Adelaide's WOMADelaide festival, where world music transforms parklands into temporary villages of global connection; London's Hyde Park summer concerts, where thousands gather under glorious trees to share in collective experience; and New York's High Line, which elevated linear park built on old rail infrastructure, where street musicians provide impromptu soundtracks to sunset walks. These moments represent what cities do best: create spaces where strangers become temporary communities through shared experience.

The music of cities is not just entertainment—it is another form of storytelling, another way of passing down knowledge and building community. It is the way rhythm and melody can teach what speeches cannot. It is how movements find their anthems and neighbourhoods find their identity. It is proof that we need more than analysis and policy to navigate transition: we need art, beauty, and collective joy.

The Work of Transition

We live in what many call a polycrisis, where systems are fragmented, authority is unclear, timelines are opaque, and people feel that decisions happen 'elsewhere.' The city is a holder of many concurrent truths. Cities demonstrate both vulnerability and resistance. They are sites of protest and suffering. Yet they also demonstrate solutions, spark imagination through creative destruction, and every now and then demonstrate that they can normalise preparedness as civic culture. They help make complexity legible without flattening it.

The Indian writer Amitav Ghosh wrote of the 'Great Derangement': our apparent incapacity to deal with wicked problems like climate change. He sees a dangerous path where individual moral action increasingly replaces collective initiatives. Thirty years earlier, *The Shadow Lines* already asked how different perspectives draw invisible lines between people—lines that connect as much as they hold apart.[2]

We needed to address this disconnect. Not with another analysis but with stories. With 20 contemporary wanderers experiencing the city and the land as a living organism—everything, everywhere, all at once. These stories do not resolve every tension that cities carry—around power and accountability, around finance and whose priorities get resourced, around whose knowledge travels and whose stays local. We did not set out to resolve them. We set out to name them honestly and to be part of the dialogues they demand.

We wrote this book because we are in transition ourselves. Because storytelling can reconnect what individualism has divided. Because we wanted to keep working together even when formal structures said we should not.

But mostly we wrote it for you. For the moments when you need to be reminded why this work matters. For the days when you want to give up. For the times when you need to know you are not alone. When you need to reject simplification, act without full certainty, and design for iteration. These stories are an invitation to find your own entry point, your own memory of connection, and your own community that needs your service—and an invitation to govern complexity together.

The land remembers. And so do we. We remember that we are part of an ecosystem, not separate from it. We remember that complexity is natural and that chaos can be beautiful. We remember that we have a choice: to be guardians for the generations that come after or to be the generation that broke the chain.

Before we close, an honest admission. Boundary spanners—ourselves included—too rarely find each other in ways that go beyond the conference and the corridor. Stuck in our own organisational logics, we do not yet fully team up at the scale our accountability demands. Often we convene without committing. We connect without covenanting. The systemic challenges cities carry will not be answered by individual

---

[2] Ghosh, A. (2017). *The great derangement*. University of Chicago Press & Ghosh, A. (1988). *The Shadow Lines*. Ravi Dayal Publishers, Delhi.

boundary spanners doing good work in parallel. They require us to genuinely build together. Walk with us. Reach out to us. The path is long, but the company is good—and good company, finally organised, is worth every step.

www.ingramcontent.com/pod-product-compliance
Lightning Source LLC
LaVergne TN
LVHW010319220926
848484LV00004BB/801
9789819595433